The Diaries of James Simmons

Paper Maker of Haslemere

1831-1868

Frontispiece: Photograph of James Simmons the diarist, probably taken by his son William on 8 August 1854, when he was 71 years old. The diary entry for that day states: 'William has been busy part of the day taking our portraits using the phrotographic art.' The original is an albumen print taken from a wet collodion plate.

(Courtesy W.J.D. Cooper)

The Diaries of James Simmons
Paper Maker of Haslemere
1831-1868

EXTRACTS SELECTED
INTRODUCED AND ANNOTATED BY
ALAN CROCKER & MARTIN KANE

Surrey Industrial History Group (SIHG)

First published by the Tabard Private Press MCMXC
© Alan Crocker and Martin Kane 1990
Wood-engraving © Michael Renton 1990
ISBN standard edition 1-869924-04-5
ISBN special edition 1-869924-05-3

Second Edition published by
Surrey Industrial History Group (SIHG)
www.sihg.org.uk

© Alan Crocker, Martin Kane and Surrey Industrial History Group
2015

ISBN 978-0-9538122-6-4

Printed by
YesPrint, 3 Leafy Oak Workshops, Cobbetts Lane,
Yateley, GU17 9LW, UK
www.yesprint.co.uk

The Surrey Industrial History Group (SIHG)
is a Group of the
Surrey Archaeological Society
Castle Arch, Guildford, GU1 3SX, UK
Company No 1160052, Charity No 272098

Contents

List of Illustrations..iv

Acknowledgements..v

Foreword..vi

Preface to the Second Edition vii

Introduction..1

The Diaries of James Simmons..21

Postscript...116

Appendices

    1 Selective Simmons Pedigree 118

    2 Chronology of the Paper Mills...................................122

    3 Structure of the Diaries 125

    4 Glossary of Technical Terms 136

    5 Making Paper on a Simmons Mould.........................140

References 143

Index of Paper Mills 146

Index of Personal Names 149

The Authors..162

Illustrations

Photograph of James Simmons..................................... frontispiece

Selective Simmons family tree...3

Table of periods covered by the diaries 6

1   Haslemere on Greenwoods' map of Surrey of 1823...................4

2   Photograph of the Simmons diaries 8

3   Vatman and coucher making paper by hand.........................14

4   Photograph of a Simmons watermark device 15

5   Ream wrapper re-used as cover..17

6   Section of sale particulars of 1832 23

7   Painting of James Simmons's wife Charlotte (colour)...............33

8   Sickle Mill from the north-east in 1850.............................38

9   An early Fourdrinier paper-making machine 50

10 Example of two facing pages from the diaries......................60

11 Transmitted light photograph of Simmons laid paper..............64

12 Sickle Mill from the west in 1850.....................................74

13 Wood engraving of Field End..78

14 New Mill from the north-east in 1850 86

15 Photograph of Mary Wornham Penfold..............................100

16 Photograph of John Wornham Penfold 101

17 Section of sale particulars of 1854 104

18 The memorial tablet in St Stephen's Church, Shottermill.......111

19 Sickle Mill House in 2015...112

20 Steam engine house behind Sickle Mill House in 2015 113

21 New Mill shortly before it was demolished.........................114

22 Sickle Mill Plaque...115

23 Photograph of James Simmons IV....................................120

24 Photograph of William Simmons......................................121

25 Simmons watermarks from paper used in the diaries I 126

26 Simmons watermarks from paper used in the diaries II.........128

27 Simmons watermarks from paper used in the diaries III.......130

28 Making paper on the Simmons mould in 1989.....................141

Acknowledgements

The opportunity to publish this volume of extracts from the diaries of James Simmons, paper-maker of Haslemere in Surrey, has arisen through the generosity of Mr W.J.D. Cooper CBE of Hindhead and Mr E.E. Orchard CBE of Haslemere. Mr Cooper has given us access to his collection of thirty-six diaries, written in the mid-nineteenth century by his great-great-grandfather, and also to other items in his family archives. In addition he has deposited the diaries at the Guildford Muniment Room*, so that they can be studied by future generations of historians interested in paper-making and local affairs. We are particularly pleased that Mr Cooper agreed to write the Foreword. Mr Orchard has deposited the two diaries which were in his possession at Haslemere Educational Museum.

Our understanding and appreciation of the diaries owes much to the frequent discussions we have had with Bob Trotter, Greta Turner and John Turner, of the Haslemere Group of the Surrey Archaeological Society. They have undertaken the mammoth task of transcribing the complete contents of the diaries and have made a major contribution to researching the lives of many of the people mentioned. Others who have helped include Robin Clarke, Shirley Corke, Glenys Crocker, Laurence Giles, Richard Muir, Cyril Queen and Violet Queen, Tim Smith and John Warren. The staff of the Surrey, Hampshire and West Sussex Record Offices, Guildford Muniment Room, Haslemere Educational Museum, Godalming Museum, Surrey Local Studies Library, Science Museum Library, Victoria & Albert Museum and St Bride Printing Library have also been very helpful. The diaries were located while one of us (Martin Kane) was working with the River Wey Trust and we are indebted to the trustees for their encouragement and support.

Two leaves of the first edition of this book are of paper made in 1989 on a James Simmons mould which is dated 1812. This mould is in the collections of Haslemere Educational Museum and we wish to thank the Curator and Trustees for giving permission for it to be used. Unfortunately, the deckle belonging to the mould has not survived and we are grateful to Bernard Oddie for making one that matches perfectly. The paper was made at Wookey Hole paper mill near Wells and we are indebted to their staff for their enthusiasm and skill with which they successfully undertook this task, which was challenging because of the delicate condition of the mould.

Finally it is a pleasure to thank Philip Kerrigan of the Tabard Private Press for many valuable discussions.

* The documents from the Guildford Muniment Room and the Surrey Local Studies Library are now at the Surrey History Centre, Woking.

Foreword

As a descendant of James Simmons III (he was my great-great-grandfather), it gives me much pleasure to write a short foreword to this book. On the death of his son James Simmons IV, who had no children, the Simmons property in Shottermill, which at that time included Cherrimans and Brookbank, came to my grandfather. It was then that the diaries on which this book is based came into the possession of the Cooper family and eventually passed through my father to me. They were obviously of local historical interest, but for some time I was uncertain what use to make of them. However a meeting with the historian Martin Kane, who subsequently introduced me to Alan Crocker, resulted in the publication of this scholarly book based on the diaries. It is most satisfactory that after all these years a useful purpose has been found for them. I trust that they will be of value to future paper historians and researchers into local history.

W.J.D. Cooper
Montana
Hindhead
25 January 1989

Preface to the Second Edition

This book was first published in 1990, when it was printed by Philip Kerrigan at the Tabard Private Press, Oxshott, as a limited edition of 210 copies. It was set in Monotype Baskerville and printed on T. H. Saunders mould-made paper. Lithographic illustrations were printed by the Senecio press, Charlbury, and line-blocks were made by G P E Ltd of Haslemere. A wood engraving was specially commissioned from Michael Renton. The book was bound by The Fine Bindery, Wellingborough. Thirty copies were quarter bound in leather and the remaining 180 copies bound in Loom State fabric. Because of the high quality of the production, copies were purchased by collectors of fine books all over the world. However, only a few copies are readily available to local and industrial historians in Surrey and neighbouring counties or to paper historians worldwide. It was therefore decided to produce a second edition to meet this need.

Since the first edition appeared further research has been carried out on the individuals who appear in the Simmons diaries and a better understanding developed on some of the aspects of nineteenth century papermaking which are discussed. It was decided, where appropriate, to include this information in the new edition without making significant modifications to the original text. It was also decided to introduce additional illustrations in this edition.

At the end of each copy of the first edition, a sheet of paper, which was made by hand in 1989 on a James Simmons mould dated 1812, was bound-in. It has not been possible to do this for the new edition but the description of how the paper was made has been retained as Appendix 5. A photograph showing a pair of sheets of this paper being made has been included.

In the first edition we were pleased to be able to thank the large number of friends and colleagues who had helped in making publication of the

Preface to the Second Edition

book possible. These acknowledgements are reprinted in full in this volume but we would like to emphasise our indebtedness to three of them here. The book would not have been possible without the generosity of William Cooper, who died in April 1994. He was the great-great-grandson of James Simmons III and gave us access to the 36 diaries and other items in his family archive. Also, Ted Orchard, who died in February 2006, allowed us to study a further two diaries that he held. Following the merger of the Surrey Record Office, the Guildford Muniment Room and the Surrey Local Studies Library, the William Cooper diaries were transferred to the Surrey History Centre, Woking. The Ted Orchard diaries remain in the archives of Haslemere Educational Museum. The other key person was Philip Kerrigan of the Tabard Private Press who printed and published the first edition. In that volume he was only willing to print a brief acknowledgement of the help he had provided. Sadly, he died in September 2012 but we would now like to state that he made outstanding contributions to the project including management, technical expertise, design, editorial thoroughness and efficiency, knowledge, research and enthusiasm.

We are also greatly indebted to several people who have helped with the production of the second edition of the book. In particular, Jan Spencer of the Surrey Industrial History Group, who, with great expertise and skill, digitised the text and illustrations from the first edition and designed the layout of the second. He provided new photographs for the book as well. Glenys Crocker also contributed photographs and helped in many discussions. It is again a pleasure to thank the staff and volunteers at Haslemere Educational Museum, particularly Kate Braun, Assistant Curator, for providing access to archive material.

Following the publication of the first edition and a full transcript of the diaries, further research by Greta Turner, one of the transcribers, led to an outstanding local history book in two volumes:

Shottermill – its Farms, Families and Mills: Part 1 Early Times to the 1700s and *Shottermill – its Farms, Families and Mills: Part 2 1730 to the Early Twentieth Century*.[67, 68]

Sadly, Greta died in July 2014.

Alan Crocker and Martin Kane, May 2015

Introduction

From 1831 until shortly before his death at the age of 84 in 1868, James Simmons, master paper-maker in the small market town of Haslemere in Surrey, kept a diary. Thirty-eight booklets in which the entries were written became available shortly before the first edition of this book was published in 1990 and provide a unique and fascinating account of the family and business affairs of a struggling country paper-maker in the mid-nineteenth century. Simmons was a prominent and well-respected leader of the community. As well as being a paper-maker he farmed over 500 acres (202 hectares) and was very much involved in local affairs in his role as church warden. The diaries reflect the diverse interests he had in work, his family, village life and also in national and international events.[1] An added interest is that the booklets are home-made and produced almost entirely from paper made by James Simmons himself. It was a difficult time for paper-makers as the traditional craft of making paper by hand was rapidly being replaced by the use of paper-making machines. These were able to produce much more paper but rags, the raw material from which all paper was made, were in short supply. In addition the available water-power was often inadequate to convert the rags into pulp so that steam engines had to be installed.[2, 3] The diaries therefore provide an absorbing insight into the concerns of a paper-maker during this crucial period of the industry. The extracts which are presented here have been selected primarily to illustrate this part of the more general story they relate.

Four generations of paper-makers with the name James Simmons did in fact operate at Haslemere,[4, 5] which is about 40 miles south-west of London in a corner of Surrey bordering on both Sussex and Hampshire.[6] To clarify their relationships, the relevant section of the Simmons family tree is given here and a more complete pedigree is provided in appendix 1.[6, 7, 8] James Simmons I founded the business in 1736 at Sickle Mill on the headwaters of the southern branch of the

River Wey, which is a tributary of the Thames. He died in 1777 and his sons James II and William became the paper-makers. Another son, Humphrey, was a London stationer and presumably sold the paper produced by his family and acted as an agent in procuring orders. James II died in 1790 and William in 1801, leaving a widow Hannah, a son James III aged 17 and five younger daughters. By this time the family were operating three paper mills, Sickle Mill and the nearby Pitfold and New or Hall's Mill. They also owned several local farms and Shotter corn mill. After William's death the paper mills were rented to John Howard until 1811 when James III, the writer of the diaries, was able to take charge. Together with his son James IV, who was born in 1815, he was active at the mills until 1849, when for a brief period John Lill Lightfoot and later Joseph Fourdrinier became tenants. Then in 1852 James IV took over the business but two years later sold Sickle Mill to Henry and Thomas Appleton. They continued to make paper there until about 1870, after which they concentrated on their main business of manufacturing military braid and worsted lace.[4, 5]

The countryside around Haslemere[6, 9] is shown in the enlarged detail of the 1823 map of Surrey by Christopher and John Greenwood reproduced as figure 1.[10] 'Sickle Mill', one mile west of the town, is shown with its large mill pond and a symbol representing a water-wheel. The boundary between Haslemere and Frensham parishes passes along the centre of the pond, which was fed by small streams rising to the north and to the south of the town. The southern stream rises near Chase Farm and forms the boundary between Surrey and Sussex, apart from a stretch of about half a mile where at that time it separated Surrey from a detached part of Hampshire. Most of the Sickle Mill buildings survive but the attractive pond has unfortunately been largely filled. Prior to the establishment of the paper mill the site was used as a corn mill and an iron forge. 'Shotter Mill' is also marked on the map being about 400 yards (366m) west of Sickle Mill. However as the buildings are south of the infant River Wey, in the Sussex parish of Linchmere, they are not shown. The mill ground corn until the 1920s when it became a builder's store. It had a leat starting near Sickle Mill and two picturesque ponds fed by springs and these survive. New Mill was 200 yards (183m) west and downstream of Shotter Mill and again the buildings were in Linchmere and are not shown on the map. It had a large pond however and one half of this and a waterwheel are

Selective Simmons Family Tree

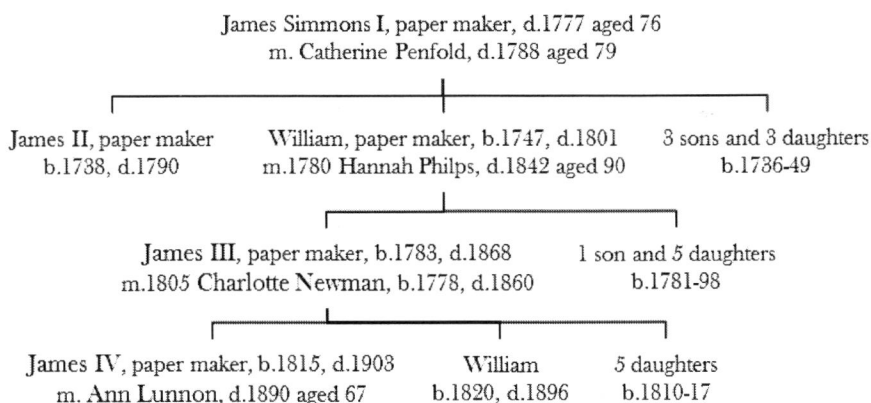

James Simmons I, paper maker, d.1777 aged 76
m. Catherine Penfold, d.1788 aged 79

James II, paper maker	William, paper maker, b.1747, d.1801	3 sons and 3 daughters
b.1738, d.1790	m.1780 Hannah Philps, d.1842 aged 90	b.1736-49

James III, paper maker, b.1783, d.1868	1 son and 5 daughters
m.1805 Charlotte Newman, b.1778, d.1860	b.1781-98

James IV, paper maker, b.1815, d.1903	William	5 daughters
m. Ann Lunnon, d.1890 aged 67	b.1820, d.1896	b.1810-17

Abbreviations: b. born; m. married; d. died

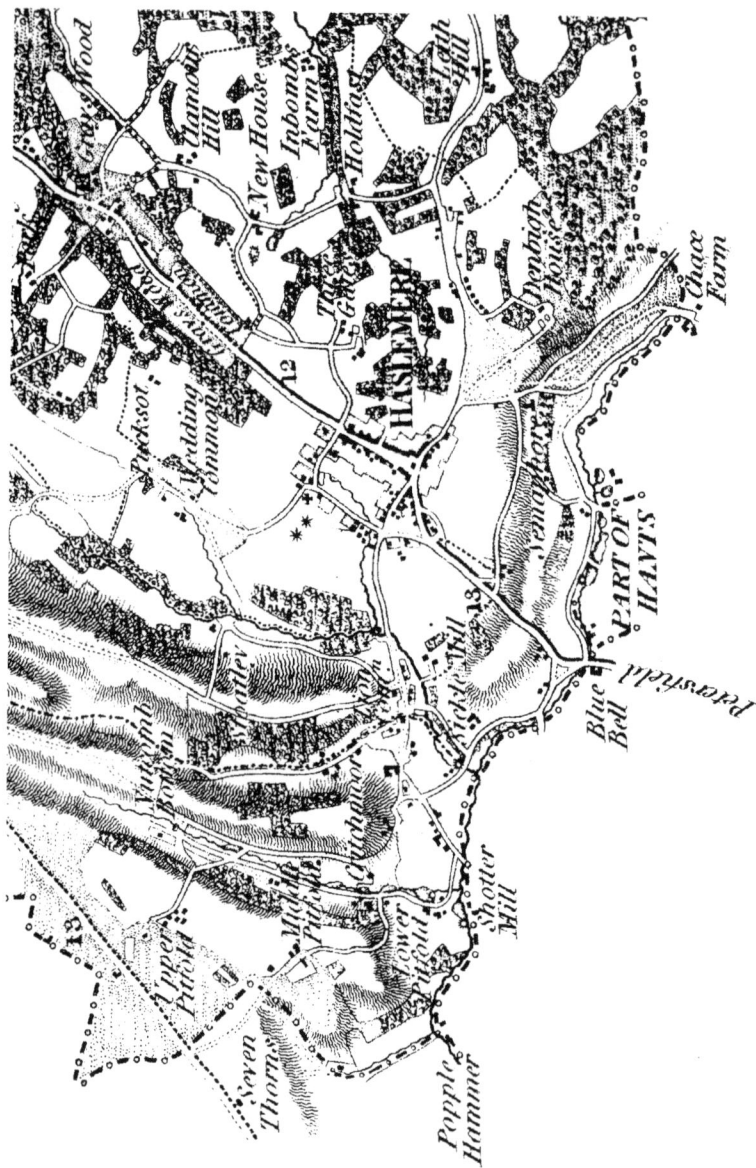

Figure 1. Enlarged detail of the 1823 map of Surrey, at a scale of one inch to one mile, by Christopher and John Greenwood, showing the neighbourhood of Haslemere in the south-west corner of the County (x 1.5). (Courtesy Surrey Archaeological Society)

indicated immediately above the letters 'Sh' of 'Shotter Mill'. After it closed as a paper mill New Mill was used for leather dressing. The buildings were demolished in 1976. Finally Pitfold Mill, in the historic Surrey parish of Frensham, was 200 yards (183m) north of New Mill on a small tributary of the Wey. The mill pond, again with a water-wheel symbol, is shown on the map to the right of 'Lower Pitfold' which was a farm belonging to James Simmons. After it closed as a paper mill Pitfold was also used to produce leather, but all the buildings have disappeared and the site has been developed for housing. A chronology of the three paper mills is given in appendix 2.

A photograph of the diaries kept by James Simmons III, which form the basis of this book, is shown as figure 2. They cover the period from 16 August 1831 to 16 January 1868, three months before their author died, and have been numbered chronologically as shown in the accompanying table. There are unfortunately five gaps in the sequence and there is nothing to suggest that diaries were not kept during these periods. Anticipating that the missing diaries may at some time become available, appropriate spaces have been left in the numerical sequence. The first and largest of these occurs after diary 1 and covers just over three years. As the average period covered by the first few booklets is about five months it is probable that eight diaries are missing. The second booklet has therefore been allocated the number 10. The next gap, after number 28, is for thirteen months and two diaries are probably missing. Then there is a one diary gap of five months after number 33. From number 36 onwards most of the diaries start on 1 January, so it is clear that the two year gap after number 40 was occupied by two diaries and the one year gap after number 49 by one. Thus there were probably fifty-two diaries in all, plus any written before number 1. In addition the collection includes a single diary kept during 1859 by James's daughter Ann and this has been included at the end of the table.

The structure of the thirty-nine available diaries is summarised in appendix 3. They contain between 64 and 250 pages, the average number being about 117 and the earlier diaries tending to be smaller. The size of the pages in millimeters ranges from 116 by 185 to 160 by 235, again the earlier diaries being smaller. Thirty-one of the diaries are sewn in one section but in the later years there are examples sewn in multiple sections including one with eleven. In these cases the pages of the different sections are not always the same size. Twenty-two of the

Periods covered by the diaries of James Simmons

1. ☼	16 August 1831	to	22 October 1831
10.	December 1834	to	30 June 1835
11.	2 July 1835	to	5 November 1835
12.	5 November 1835	to	23 April 1836
13.	24 April 1836	to	13 August 1836
14.	14 August 1836	to	24 December 1836
15.	25 December 1836	to	25 August 1837
16.	27 August 1837	to	22 February 1838
17.	23 February 1838	to	10 October 1838
18.	11 October 1838	to	12 October 1839
19.	13 October 1839	to	12 March 1840
20.	23 March 1840	to	14 March 1841
21.	20 March 1841	to	2 September 1841
22.	3 September 1841	to	31 March 1 842
23.	31 March 1842	to	18 September 1842
24.	19 September 1842	to	22 May 1843
25.	24 May 1843	to	11 February 1844
26.	12 February 1844	to	15 December 1844
27.	21 December 1844	to	4 August 1845
28. ☼	8 August 1845	to	24 March 1846
31.	April 1847	to	26 September 1847
32.	3 October 1847	to	22 April 1848
33. ☼	23 April 1848	to	19 November 1848
35.	April 1849	to	31 December 1849
36.	1 January 1850	to	31 December 1850
37.	1 January 1851	to	31 December 1851
38.	1 January 1852	to	31 December 1852
39.	1 January 1853	to	31 December 1853
40. ☼	1 January 1854	to	31 December 1854
43.	January 1857	to	15 December 1857
44.	1 January 1858	to	31 December 1859
45.	1 January 1860	to	31 December 1860
46.	1 January 1861	to	31 December 1861
47.	26 January 1862	to	31 December 1862
48.	1 January 1863	to	31 December 1863
49. ☼	1 January 1864	to	31 December 1864
51.	1 January 1866	to	30 July 1867
52.	31 July 1867	to	16 January 1868
AS	10 June 1859	to	31 October 1859

☼ Gap in sequence

AS = Diary of
Ann Simmons

The diaries have been
deposited at Surrey
History Centre except
for 36 and 37 which
are held by Haslemere
Educational Museum.

diaries have no covers. The remainder have covers usually formed by gluing a coarser piece of paper to the outer pages of the booklet.

The entries in the diaries were written by James III on both sides of the paper, except when pages were missed accidentally or when gaps were left deliberately with the unfulfilled intention of inserting information later. The period covered is written on the cover. Twenty-two of the diaries have an index inside the front cover and in some cases the back cover is used to complete entries. Almost all of the writing is in ink, most of which appears brown but occasionally is black. Most pages contain about 120 words, the writing in the later diaries which have larger pages also being larger. Some of the entries in the last few diaries have very large writing giving about 60 words to the page. Simmons presumably wrote these without his spectacles. Entries were made on about one-half of the available days but when there is a gap the next entry often summarises information for the missing period. There is almost invariably an entry for each Sunday, which is usually longer than those for weekdays. With one exception the number of pages written each month ranges between 10 and 18, which seems remarkably consistent. The exception is diary 1 with 40 pages per month, which suggests that this may truly be the first in the series, being written in a particularly conscientious manner. It is also interesting that the average number of pages dropped from about 16 to 12 per month when Simmons effectively retired in 1849.

The content and character of the entries in the diaries can be demonstrated most effectively by means of quotations. The major part of this volume contains all the extracts recognised as referring to paper-making. However, in order to indicate the breadth of information covered by the diaries, to illustrate different aspects of Simmons's personality and to place his paper-making activities in context, a brief selection of more general extracts, in most cases in abbreviated form, will now be presented.

As a large proportion of the diary entries concern James III's views on religion it is appropriate to commence with a few of these:

> 20 August 1831. The week passed very favourably for the harvest - the Lord is very merciful and is my help & my shield. In looking back through my past life I find vain & foolish thoughts to have been my besetting sin.

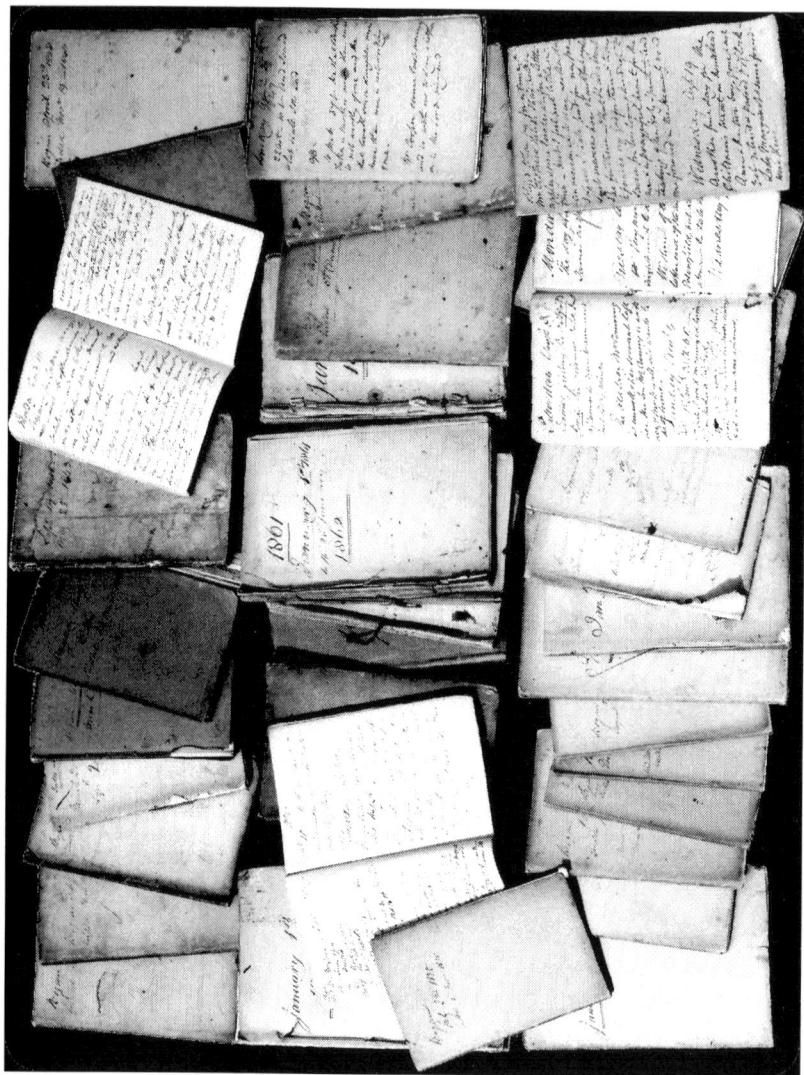

Figure 2. The diaries of James Simmons. (Photograph by Gareth Crocker)

28 November 1840. Our Wedding day – 35 years past. 5 years of sabbaths have we lived together.

30 August 1848. We have heard for some time that Mrs Tilbury has turned sceptic and do not believe the Bible.

Simmons was very involved in the local church and was a key figure in establishing a new church at Shottermill:

1 July 1835. Gave the children of the Sunday School their usual treat of Cake & wine. We had more than 100 children met together.

7 April 1838. The smallpox continues. We had but few children at the Sunday School on account of it.

29 March 1842. The people very busy to get the Church ready. We are also very busy as the Clergy meet & lunch at Sickle Mill.

31 March 1842. The opening of the New Church at Shotter Mill to be called St Stephens. The Archdeacon [Mr Wilberforce] did not arrive till after the service was began. Mr Bowles was to have brought him from Godalming, did not wait, thinking he would not come. He walked to Witley and at last met with a farmer who drove him to Shotter Mill. The Church was exceedingly full one person said 420 another 375.

25 March 1845. Went to the meeting at Frensham Church. I am the overseer for this end but as Dutton is continued assistant overseer, it is almost a nominal office.

2 May 1848. Mr Candy had a letter to set out the site for the [Parsonage] House – the Bishop appointing me Surveyor.

In addition to being a paper-maker Simmons was a farmer :

22 October 1831. Very much engaged in worldly concerns. Made about a pipe of Cyder, sowed Wheat, dipped the lambs, weighed the wool. Trade dull.

24 February 1835. Went to Worplesdon to a sale at a nursery ground. Bought a lot of apple trees – five or six hundred for 26s.

9 March 1837. I sow'd my oats in the Lion field – rolled my wheat & seeds &c. Draining in the Sturt Meadow piece under & upon the brow in front of Sickle Mill.

20 May 1837. Cut 29 oak trees in the New Mill Hanger – got them out and planted the moor piece with potatoes.

10 May 1838. This week two of my calves bellowed out, ran round the pen appeared in great misery and died. A man called and said it was occasioned by the cows eating grass. I think myself it was fulness of blood on the brain.

21 June 1845. Took up my little clover hay in good order. Thatched the rick today.

19 April 1852. We lost some eggs from the yard and have the policeman to watch – he saw the man who works for Mr Fourdrinier come into the yard and took 4 eggs and put them into his pocket.

11 October 1852. The sale of my Farming Stock today which was a fine day for it. The Cattle fetched quite as much as I could expect. The implements went low but they were old. The Hay did not sell.

1 August 1853. The Meadows so long cut now clearing off the Hay of no value. I can scarcely remember a year in which so much has been spoiled. The harvest very backward and the wheat crop very bad.

25 October 1864. A great crop of Grapes & they are tolerably ripe.

The diaries throw light on the cultural and leisure activities in which the Simmons family became involved:

14 October 1839. Went to the Town Hall in the evening to hear a lecture on Astronomy by Mr Edmonds a dissenting minister from Petworth.

23 June 1841. Started this morning for Windsor. We drove through the Park down the long walk, we had some lunch & went to the Castle, we walked on the terrace and then went up the round tower, the view is splendid and the building magnificent.

5 May 1842. Mrs S. went to see the Wax work in Baker Street.

8 May 1854. Went to the British Museum – went over all the Rooms a most wonderful collection, after looking round on the Birds, Beasts, Fishes, Insects &c., the Minerals, Egyptian & other antiquities – well may we say "Wonderful were thy works O Lord in wisdom hast thou made them all." I want to go once more as I have not seen the Library which I understand contains two Millions of Volumes.

21 April 1857. Went this evening with John [Cooper] to the Town Hall Haslemere to hear Mr Newman the Ventriloquist, my little companion was very much amused but I cannot say I was, for his performance & power was not of the first order and the subject matter of it very trifling.

28 December 1861. Catherine, Ann & Myself went to the South Kensington Museum. Very well worth seeing.

Various sporting activities are also mentioned:

28 August 1848. I yesterday [Sunday] spoke to the young men who were playing Cricket on the Common – they persisted in it and were very rude. I must use the little means I have to prevent it and leave it in the hands of the Almighty.

7 January 1854. William left us this evening. While here he killed a good many Blackbirds, Fieldfares, &c.

22 April 1857. Mr Moline & Mr Kidd called to have a days fishing – they caught a few trout and for a little time I was with them. I caught some perch with a worm – two or three very good ones.

12 June 1859. [Diary of Ann Simmons] No cricket on the Common tonight. Last Sunday Papa went out and spoke to them on the evil of it. My beloved Fathers quiet influence and testimony against evil does I believe tell in this place.

Many entries are concerned with his own health and that of his family and friends:

15 June 1839. I have been in continual pain from a gathering in my jaw bone. I have kept it continually fomented with hot water outside and within – and this afternoon it broke and I am very much relieved. I could not open my mouth, neither could I sleep at night. But the Lord has relieved me.

31 December 1839. Arthur Newmans eye continues in a very doubtful state. He is continually having leaches.

20 May 1840. James gone to Chichester today, the children has had the measles.

2 June 1852. My birthday – 69 years old. Saved to this time by the providential mercy of God through many dangers. 1st by the falling of a Horse by which my arm was broken. 2nd by a blow when drawing out some timber which broke my leg and put out my

shoulder. 3rd by a severe strain in my ankle in getting over a hedge. 4th by being thrown out of the chaise when the pony ran off and putting my thumb out of joint – and lastly by a gun falling and going off by which the charge went between my thumb and forefinger but which only slightly wounded them.

The later diaries indicate the problems he was having with old age:

2 June 1864. My birthday. Ann put into my room a large printed testament that I might read it without spectacles.

25 December 1867. Christmas Day, dear Ann with myself spent it alone our servant Rhoda gone home a holiday till tomorrow. My "rash" still being very troublesome do not get much sleep. Mr Clothier's medicine &c. helping on.

The diaries also contain descriptions of natural phenomena:

15 May 1836. The service in the evening, because of an eclipse of the sun in the afternoon. Many, many eyes in England (mine was one of them) lifted up to see the sun darkened in the firmament.

1 February 1842. Today very fine and mild. I observed a Bat flying over the New Mill pond this afternoon picking off flies.

2 October 1858. The Comet very brilliant the last night or two; it is to be seen to the north of Arcturus, the tail more splendid to me than any one of them I have seen before. I remember the one in 1811. "How wonderful are thy works O God."

Other entries refer to public health and disease:

17 January 1840. Very Dirty in London. Observed a piece of the pitching in the main road opposite White Hall to be made of wood & although surrounded on three sides by dirt & slop on the pitching this was comparatively dry – no slop and very much less noise.

13 September 1849. The Cholera still prevails in London – it is rather increased I hear last week the deaths being 380.

National events are occasionally reported:

28 June 1838. This day Queen Victoria was crowned; it was kept in London as a holiday, here it was not. We gave the Sunday school children a treat on the front Green at Sickle Mill & the teachers also. We then went to Haslemere Church – our boys went to see some fireworks let off.

27 February 1858. I have been today to get signatures against the passing of the bill for the admission of Jews to Parliament.

15 December 1861. On going to Church this afternoon heard the distressing intelligence of the death of the Prince Consort.

5 February 1867. England is falling away from her high standing she has hitherto held among the nations of the earth and her power will lesson before the coming of Christ.

Finally there are reports of major international events:

23 September 1854. The Papers report this morning that the combined armies are landed in the Crimea – 25000 English, 20000 French & 12000 Turks without opposition are moving forward to Sebastapol. The Lord has so far helped us, may the prayers of God's people be continually ascending to the throne of grace. The cholera is on the decrease.

28 May 1864. Terrible doing in America, there has been fighting for the last week or more – report say forty thousand killed & taken prisoners.!!!

In order to appreciate fully the extracts from the diaries concerning papermaking an understanding of the main technical processes involved is helpful.[2, 3]

An outline of the methods used is therefore given here and a glossary of technical terms is presented in appendix 4. Until about 1860 almost all paper was made from rags, which had to be collected, sorted, cut into small pieces, washed and left to ferment in water. White linen or cotton rags were used for writing paper and coarse rags for brown paper. However chlorine was discovered in 1774 and enabled coloured rags to be bleached and therefore used for white paper. The fermented rags were pounded, churned and macerated by water powered hammers or, from about 1700, in an engine called a Hollander or beater to produce a creamy pulp known as stuff. This was then passed through a knotter to remove unwanted lumps into a vat where it was heated and agitated to keep the fibres in suspension. Individual sheets of paper were made, as shown in figure 3, by a vatman who dipped a mould consisting of a rectangular wooden frame with a cover of wire mesh into the stuff. He lifted it out horizontally with the mesh supporting a layer of matted fibres. The edge of this layer was defined by a wooden frame called a deckle which was placed around the mould

Figure 3. Vatman and coucher working together making hand-made paper. The vatman on the right is about to dip a mould, fitted with a deckle, into the stuff. He will then lift it out horizontally supporting a wet sheet of paper, remove the deckle, and place the mould on the bridge across the vat. The coucher on the left is placing a piece of felt on a sheet of paper at the top of a stack of interleaved sheets of paper and felt. He will then take the second mould, resting against an asp on the bridge, turn it over depositing the wet paper on the felt and return the empty mould to the bridge. The vatman now takes this mould, places the deckle around it and forms the next sheet of paper. Meanwhile the coucher takes the full mould from the bridge, rests it against the asp and places the next felt on his stack. (Wood-cut from the *Penny Magazine*, 1833,[38] based on a drawing made at Turkey Mill near Maidstone, Kent)

Figure 4. Photograph by University of Surrey of one of the Post Horn
watermark devices sewn on to a laid cover of a mould which also has
two J SIMMONS 1812 countermarks. The stitches attaching the cover
to the wooden ribs of the mould, which are behind the vertical chain
wires, are visible at the left. The mould measures 1030 by 424mm
which is consistent with making two sheets of post printing paper.
The chain and laid wires have spacings of 24 and 1.2mm respectively.
The mould-maker was John Green, jnr., of Maidstone.
(Courtesy Haslemere Educational Museum)

before dipping it in the vat. The wet sheet of paper was turned over on to a piece of felt by a second workman called a coucher who built up a stack of sheets of paper interleaved with felt. This post of paper was placed in a wet press to squeeze out as much water as possible and a third man, the layer, then separated the sheets of paper and hung them over ropes in a loft to dry.

The watermark in the paper was formed by a design fixed to the wire mesh of the mould.[2, 3] This mesh could be either 'laid' or 'wove'. The laid mesh consisted of parallel fine wires about 1mm apart fixed to a perpendicular set of wires about 25mm apart. These wires leave a characteristic pattern of laid and chain lines respectively in the paper. A photograph of part of the laid cover of a mould used by James Simmons is shown in figure 4.[11] The wove mesh is made of finely woven wire and leaves a very indistinct impression resembling fabric in the paper. When the paper was dry it was flattened in another press and glazed by polishing with a stone, hammering, passing through rollers or pressing between plates in a large finishing room called the salle or sol. Paper for writing also had to be sized and this was done either by mixing size with the pulp or dipping the dry sheets into a tub of size. China clay could be introduced into the stuff to make the paper whiter and heavier.[2, 3] Finally the paper was wrapped in packets and despatched to the stationers. However before this could be done it had to be inspected by an exciseman as until 1861 paper was taxed.[12, 13]

After the first edition of this book was published in 1990, a remarkable discovery was made regarding the paper being made by James Simmons III shortly after he became the master paper-maker in 1811.[64] The cover of a booklet containing the land tax records for Haslemere in 1820 is made of coarse blue paper printed on the inside with the Royal Arms and the words 'SUPERFINE FINE / JAMES SIMMONS' as shown in figure 5. Also the pages of the booklet have Britannia with 'JAMES SIMMONS / 1814' watermarks. The word 'Superfine' clearly indicates that James Simmons claimed that his paper was of exceptionally good quality and the Royal Arms includes the German arms ensigned at the centre with the Electoral Bonnet that was formally only employed between 1801 and 1816. The two small torn pieces of white paper with superimposed lettering seen in figure 5

Figure 5. Ream Wrapper.
Detail of the inside front and back covers, placed contiguously, of the
cover of the booklet containing the 1820 Haslemere land tax assessments.
(Courtesy Surrey History Centre, QS6/7/115-116)

indicates that the blue paper was originally used as part of a wrapper for a ream of 480 sheet of paper. The full version of the printed lettering would have been. 'No. of Mill: 118: lbs. / FIRST CLASS' and '14½' has been added by hand before 'lbs'. The mill number was the excise number introduced in 1816 and 14½lbs was the weight of the paper. Also a stamp has been used partly on the white paper and partly on the blue paper to impress 'CH … ON / MAKER / 815' in two places. This indicates that excise duty had been paid on the paper, '815' being the excise officer's number. This section of a ream wrapper, still displaying its excise label and stamps is the only example known to survive from the early nineteenth century, presumably because purchasers were instructed to destroy them so that they could not be re-used on other reams of paper in order to avoid paying excise duty.[65]

At the end of the eighteenth century a machine was invented which enabled a continuous roll of paper rather than individual sheets to be made. This was developed by Bryan Donkin who set up a factory in Bermondsey using finances provided by Henry and Sealy Fourdrinier who were London stationers. It consisted of a vat from which stuff flowed over an apron on to a continuous moving web of wire and thence around drying rollers covered with felt and pressing and glazing rollers heated by steam before being reeled as paper. The reels were then fed into a cutting machine to produce stacks of individual sheets of paper for writing or printing. As the moving web was woven the early Fourdrinier machines produced paper which resembled hand-made wove paper. However a technique of impressing a laid pattern and watermark on the paper was soon developed. The machine itself did not require much power but as it could produce paper much faster than a vatman more pulp and hence more beating engines were needed. These engines required a substantial amount of power so that machine mills often had to install steam engines to supplement the power provided by their waterwheels.[2, 3]

The shortage of rags for paper-making became increasingly worse until in about 1860 the first successful alternative was discovered. This was esparto grass which was imported from Spain and North Africa and made excellent paper. Then methods of producing pulp from timber were developed and today almost all paper is made from wood pulp using machines based on the early Fourdrinier models. The last commercial hand-made paper mill in Britain closed in 1987.

Returning to the diaries, the paper from which they are produced is of three distinct types. There are about 320 sheets of hand-made laid, 380 of hand-made wove and 490 of machine-made paper. The laid paper is white, blue-white or cream, the wove is cream with a few grey sheets and the machine-made is cream. Each type ranges from about 0.06 to 0.23mm in thickness. The early diaries are of hand-made paper, diaries 28 to 40 are machine-made and the later ones are a mixture. The hand-made paper has watermarks showing that all but about 50 sheets was made by James Simmons III himself. Most contain dates showing that the paper was made between 1817 and 1841. About 25 pages are of paper containing a watermark portraying the Royal Arms as used between 1714 and 1801 and this was probably made by James I or one of his sons. Also of interest is the watermark contained in the paper of diary 23 representing the arms of The United Ionian Islands. This paper is mentioned in the diary entry for 11 May 1840. The machine-made paper has no watermarks but was almost certainly made by James III after he purchased a second-hand machine in 1840. The later diaries with a mixture of papers correspond to the period after 1854 when James IV sold Sickle Mill. The diarist was clearly using his stock of old paper to make the booklets. The paper not made by James III bears the watermarks of Shepheard & Sutton, Charles Skipper & East or T Edmonds. It is interesting that one of the Edmonds sheets also has the countermark 'Not Bleached', indicating that bleached paper was considered to have inferior qualities.[14] The covers are of buff, grey-buff, light brown, pink or mid-blue wove paper, of grey or grey-brown laid paper or of cream machine-made paper. This paper ranges up to 0.33mm in thickness.

The paper used in the diaries demonstrates that James Simmons made a wide range of high quality papers and the entries provide further information about his products. Thus he refers in general terms to making bank note paper (11-13 June 1836), wrappers by hand (27 February 1835) and by machine (22 March 1841), large paper (12 November 1842) and tissue (2 December 1852), at least some of which was pink (9 February 1853). Other entries refer more specifically to foolscap (2 July 1835, 11 May 1840), small hand (27 March 1841), lumber hand (2 April 1841), demy (6 April 1841), news demy (2 October 1841) and plate demy (22 January 1844). These indicate the size and quality of the papers as explained in appendix 4. Unfortunately no trade catalogue of Simmons's paper has been discovered and it seems that he was prepared to make any paper for which an order could be obtained.

In recording an account of his life to be passed on to future generations of his family, it appears that James Simmons intended to provide 'moral guidance through example'. This factor no doubt influenced what he chose to write and indeed references to his paper-making concern appear only infrequently in the diaries. Nevertheless, because of the paucity of other contemporary accounts, these entries provide a significant contribution to our understanding of the life of a paper-maker during a difficult period of the nineteenth century. It is most fortunate that the diaries have been cared for by several generations of James Simmons's descendants and that their contents can now be made generally available.

The Diaries of James Simmons

This selection of extracts from the diaries, which is reproduced here in chronological order, is concerned primarily with paper-making. In most cases entries have been abbreviated but as far as possible the original spelling, punctuation and use of capital letters has been retained. In the originals the day of the week and the full date, for example 'Tuesday 30th. August', are given for each entry. Here, this information is provided only for the first extract of each month. Many of the entries have been annotated and further information is provided in the appendices and indexes. A full transcript of the diaries has been prepared and reference copies are available at the Haslemere Educational Museum and at the Surrey History Centre, Woking.

1831

AUGUST

30. [Tuesday] Mr Everitt came in the morning and told he would leave the Mill as trade was so bad he could not keep on and lose all he had – he said the rent was too high. He had before entered into an agreement with me and all was settled, and I had sent him the draft to look over.

> Samuel Philip Everitt was the tenant paper-maker at Pitfold Mill. He was still at the mill in July 1832 when, together with New Mill, Shotter Mill and several farms and cottages, James Simmons offered it unsuccessfully for sale. A part of the title page of the sale particulars[15] is shown as figure 6. Pitfold Mill was described as a substantial paper mill suitable for making the finest paper, having one vat, a waterwheel 17 feet (5.2m) high, an engine, chests and excellent spring water for washing. The rent was £70 per annum and Everitt was said to be a yearly tenant. The diaries reveal that Everitt had left Pitfold by January 1835.

SEPTEMBER

1. [Thursday] I have heard Mr Everitt intends leaving the Mill immediately without giving me regular notice.

2. Went to Godalming and to Catteshall Mill to give Mr Sweetapple my opinion respecting the value of it – they have been severely afflicted in temporal concerns.

> Thomas Sweetapple was the master paper-maker at Catteshall Mill, Godalming. He was declared bankrupt in December 1830 but continued as the tenant at the mill until 1865.[16]

7. Went to Mr Everitt – both him his wife more composed and said they did not wish to leave the Mill if they could get a living, that they would stop till the spring and try and I gave them leave to use the drying house at the Lower Mill.

> 'Lower Mill' must mean New Mill, which had a less sheltered location so that the drying house would be more effective.

9. Busy this morning putting in a new nut to the vatt press of the first side.

10. Placed the press up again, putting it together first. Raised it up some height when the pulley block broke and let it down again – providentially no one was hurt.

1835

JANUARY

10. [Saturday] The last two days I have been very busy with the Men – cutting the stream straiter below the New Mill – taking a piece out of Linchmere and giving it back just above so as to make it as nearly equal as possible.

> When New Mill, also known as Hall's Mill, was offered unsuccessfully for sale in 1832[15] (see figure 6), it was described as a compact substantial mill, driven by a powerful stream of water, with one vat and bleaching, sizing and rag houses. The rent was estimated to be £70 per annum. The fall of water was upwards of 5 feet (1.5m) which could be increased by sinking the water course below the mill. Simmons is now achieving this by straightening the water channel. The River Wey was here the boundary between Frensham and Linchmere parishes.

Particulars

OF SEVEN VALUABLE FREEHOLD & COPYHOLD

Farms,

With convenient Dwelling Houses, Cottages, Yards, Gardens, and Agricultural Buildings,

TWO PAPER & ONE CORN MILLS,

With Engines, Vats, Geer, Drying and Store Houses,

A commodious Dwelling House,

A PLEASANT COTTAGE RESIDENCE, WITH MEADOW LAND ATTACHED; ORCHARDS, FISH PONDS, PLANTATIONS,

And an abundant Supply of Water,

CONTAINING TOGETHER

Upwards of 520 Acres

Of good Meadow, Arable, Pasture, Coppice Land, and Fir Plantations.

The Estate is well stocked with thriving Oak Timber, and abundantly supplied with Fish and Game, lies very compact, is let to respectable Tenants; and is situated in the several adjoining Parishes of

Haslemere, Farnhurst, Frensham, and Linchmere, in Surrey and Sussex:

Which will be Sold by Auction,

By Mr. PEACOCK,

AT THE WHITE HART INN, GUILDFORD,

On Saturday, July 14, 1832, at 3 o'Clock, in 11 Lots,

Figure 6. Part of the title page of sale particulars of 1832 (x 0.72) when James Simmons, unsuccessfully, offered much of his property, including Pitfold and New Mills, for sale. Further details are given in the notes on the extracts dated 30 August 1831 and 10 January 1835. (Courtesy Farnham Museum)

23

17. Rode over to Mr Warren about Glue Prices and called also at Mr West's the mill below.

> William Warren was the master paper-maker at Bramshott Mill from 1822 to 1861 and James West at Standford Mill from 1832 to 1836. Warren also acquired Barford Lower Mill in 1837 and Standford Mill in 1842.[5] Paper-makers used glue for sizing their paper.

FEBRUARY

2. [Monday] Preparing for setting on the New Mill.

3. Mr Pewtress called.

> Thomas and Benjamin Pewtress and their descendants were master paper-makers at Iping Mill from 1826 until about 1870,[17] at Eashing Mill from 1833 to 1868, at Grange Mill, Bermondsey, from 1839 to 1842 and at Stoke Mill, Guildford, from 1852 to 1863.[5] They are also reported to have been at New Mill in 1839,[18] but the diaries do not confirm this. However for a brief period in 1850 they were at both Sickle and New Mills (see 10 August 1850).

7. Busy at the New Mill. Began today boiling Rags at Pitfold.

14. Yesterday & today busy at the New Mill. Beat off the first Engine of half stuff this evening.

21. This week has past very busily what with settling about setting on the New Mill, beating off the stuff, Bleaching, taking Stock &c.

27. Began making wrappers at the New Mill, James, Jno Harding & Jas Harding the set.

> The 'set' refers to the team of three, the vatman, the coucher and the layer, needed to make paper by hand.[2] The procedure is illustrated in figure 3 and explained in the caption. 'James' is James IV and several generations of the Harding family worked for James III. They were perhaps descended from Abraham Harding, tenant paper-maker at Barford Upper and Lower Mills in the late eighteenth century.[5] Later two members of the family emigrated to Middletown, Ohio (see 14 December 1850), and founded the Excello and Franklin paper mills.[19]

MARCH

9. [Monday] Set on the New Mill.

21. Mr Ward came (a gentleman from Beccles in Suffolk) a grocer who collects a very considerable quantity of Rags – his price was too high for me. I am to have a small parcel as sample.

> This is the first of many references to rag dealers, rag collecting and the difficulties of obtaining rags.

APRIL

9. [Friday] Busy putting the new fly wheel at Pitfold.

10. The Millwrights are at Pitfold putting in the fly wheel &c.

15. Called on Mr Appleton whose son came down to look at Pitfold Mill – to put up some tackle for spinning & weaving.

> The Appleton family made braid and epaulettes for military uniforms and gave its name to a company that still manufactures worsted embroidery thread. They were at Pitfold Mill from 1835 to 1837, at Elstead Mill from 1837 to 1880 and purchased Sickle Mill from James IV in 1854 staying there until about 1920.[5, 20]

18. Mr Appleton & his son came about Pitfold Mill, we looked it over and I am to hear again in a very short time.

23. Got off the first Engine of half stuff at Pitfold. Two racks for draining it finished &c.

24. Heard from Mr Appleton this morning about Pitfold Mill.

30. Mr Appleton came this morning and settled about Pitfold and the alterations to be made.

MAY

1. [Friday] Sent the agreement to Haslemere for Mr Appleton.

8. Went with Mrs Simmons to Godalming. I had a look over Mr Roakes Mill to see the work was done according to contract – & so far as I could judge I thought it was done very well.

22 & 23. Very busy at Pitfold Mill making the alterations required, for Mr Appleton to put up his tackle. Pulled the Vatt press down being very much decayed, cut a door way out of the sizeing House.

JUNE

30. [Tuesday] Elias Puttick & Isaac Salmon bound apprentices. Their yearly pay to commence from this day.

JULY

2. [Thursday] Went to London on the Guildford coach – engaged to make 500 Rm Fcap [reams of foolscap] for Messrs Hickman & Marriott and 500 Rm for Messrs Venables – the former for the Stationery Office, the latter for the East India Company.

> James III and later James IV went to London about once a month mainly to visit wholesale stationers to get orders for paper. Messrs Venables was a particularly important family firm controlled at this time by William Venables, an Alderman of the City of London from 1821 until his death in 1840.[13] It also operated paper mills in Buckinghamshire and opened Woking Mill in 1840.[5] Occasionally other members of the Simmons family visited London and stayed with relatives and friends in, for example, Vauxhall and Lambeth.

4. Mr Appleton Senr came down to Haslemere. They begin to do a little work at Pitfold Mill, but do not wholly get on.

6. Fished New Mill Pond some good trout & Eels. Employed about nine men & boys in mudding the pond. I am trying to fill up a crook in the stream below New Mill with the mud, having cut the stream more strait.

7. Succeeded very well in filling up the hole with mud. Settled with Mr Appleton about taking away the Vatt at Pitfold. If he continues to keep it for three years at the rent now let at I am to be paid nothing, if he leaves before that time I am to be paid ten pounds towards replacing it.

10. Went in the afternoon to Godalming to settle with Mr Stedman and arrange about the Insurance. Yesterday we were moving stuff to the New Mill.

11. Heard this day that the Iping Mills are burnt down and I fear it is true. I hope and have little doubt that they are insured.

13. James went to Iping – the fire did not do so much damage as was expected. Mr Pewtress is of opinion that it was not wilfully done.

AUGUST

8. [Saturday] The weather so exceedingly dry that we only made 7 days on one side & 8 days at the other Vatt during the last fortnight. Very little more than one Vatts work. I never remember anything like it before.

> Many entries in the diaries commence with a comment on the weather, which appears to have been more extreme than it is today. The weather was clearly important for the Simmons farming activities and also for paper-making. In this example there was not enough water to power the waterwheels regularly.

SEPTEMBER

19. [Saturday] Business bear heavy upon me. Trade dull – in some difficulty about some Fcap [foolscap] made for the Stationery office, part not quite sound & then all is rejected.

OCTOBER

13. [Tuesday] Went to Godalming about the small copper in the sizing House which had given out – could not get it mended.

16. Went this afternoon to Godalming and as I purposed to put the smaller copper within the larger to boil the size, I ordered a new bottom to be put in.

17. The water is more plentiful. Did at Sickle Mill 11½ Days & at New Mill 11¾ Days the last fortnight – before, for some time not more than 7 or 8 Days a fortnight.

NOVEMBER

14. [Saturday] Put up a Stove in the Rag House at Pitfold so as to make the room above hot for hardening Paper.

1836

JANUARY

28. [Thursday] Attended a meeting of the Master Paper Makers at the London Coffee House. Mr Dickenson in the chair – It was resolved to send up a deputation to the Chancellor of the Exchequer to state to him the necessity of lowering the duty on paper one half and also to petition the House of Commons to the same effect. I am not of opinion it will have any effect.

> John Dickinson was a London wholesale stationer who started his business in 1804. He also had mills in Hertfordshire and invented amongst other equipment the 'Cylinder' paper-making machine, which was a rival to the Fourdrinier machine developed by Bryan Donkin.[2] He played a leading role in the campaign for the penny post and the authorisation of stationery envelopes which he then manufactured. The firm named after him still produces stationery including the well-known Croxley brand.[21] Since 1803 the duty on first and second class paper had been 3d. and 1½d. per lb. respectively.[12]

FEBRUARY

12. [Friday] Started this morning for Winchester called Mr Warren and went to Alton. Mr Spicer went with us. It was a meeting of the trade at the Black Swan – only Mr Gater and Mr Brookman came – we talked over the high rate of wages & agreed to meet again in the Autumn of the year to effect a reduction, and also thought that 25s/[cwt] was sufficient to give for fine Rags. The next morning I walk'd round the Cathedral & the City – it appears rather a dull place – no manufacturing in it.

> John Edward Spicer snr. moved from Buckinghamshire to Alton in 1796 and his son John Edward jnr. was born in 1803. The mill at Alton had a tenant paper-maker until 1826 when the Spicers took over. The father died in 1845 and the son in 1869 but their descendants continued to make hand-made paper at Alton until 1909 when they moved to a new machine-mill at Eynsford in Kent.[5, 22] John or Edward Gater and William Brookman were master paper-makers at Up Mill, South Stoneham and at Test Mill, Romsey in Hampshire.[23] James III was clearly excited by industry and machinery but does not appear to have appreciated architecture.

19. Mr Pewtress called and drank tea.

MARCH

10. [Thursday] The weather having been unfavourable for drying & finishing paper & having a considerable parcel of a special paper on hand which I was disappointed in the sale of I have been short of cash for some little time and uncomfortable on that account but this week I have sold some of the paper before mentioned & have had paid to me some cash for different small parcels of paper goods that I am relieved in my mind on that account.

> In winter there was plenty of water power but the weather was often too cold and wet to dry the paper after it was made.

12. Busy today preparing to put in a new trough over the south water wheel at Sickle Mill & also a new Pitt wheel. It is not yet known if anything will be done about the duty on paper – I think trade would now be brisk if that was not in contemplation.

> The trough carried water to the top of the waterwheel indicating that it was overshot or pitch back.

24. Revd Hy Baker called & bought Lane End & Pitfold for 4200£.

> Lane End and Pitfold were two of the farms which Simmons offered for sale unsuccessfully in 1832 (see figure 6).[15]

APRIL

13. [Wednesday] The question of reduction of duty not being settled the trade is in an unpleasant state, there is a disposition to buy, yet the stationers hold off on account of the uncertainty of what will be the result.

23. I heard this morning that the Duty on Paper is not likely to be altered – only that Mill boards are to be the same as Brown Paper 14s/ pCwt.

> From 1816 to 1836 the duty on millboard, which was pasteboard made from first class paper, was £1. 8s. per cwt.[12] A reduction to 14s. per cwt. or 1½d. per lb. would make it the same as for brown paper.

MAY

9. [Monday] From letter received I find it is the Chancellor intention to take of half the paper Duty leaving 1½ plb [1½d. per lb.] also the duty on stained papers, and reduce the newspaper stamp duties to one penny. How this will effect our trade I cannot say. I do not see it will do any harm & probably it may do good.

19. Mr Jas Baker and Mr Jas Parson called and signed the agreement for the sale of Lane End & Pitfold.

23. Busy with the Millwrights & Plumbers.

JUNE

11. [Saturday] Began beating some new Pieces for Bank note paper. The first I have made. I do it in fear lest I should fail.

> Several local paper-makers made bank note paper including William Warren at Bramshott in 1828 for the Bank of Brazil,[24] Charles Ball and his family at Albury Park and Postford before they became bankrupt in 1820-21[25] and Portals at Laverstoke, where Bank of England paper has been made since 1725.[26] It is not known for which bank James III was making paper.

13. Began making the Bank Note paper. I think we succeed better than I expected.

15. The Bank Note paper continuing to go well.

28. Settled the business of the sale & I received the money.

JULY

9. [Saturday] Arose early and arrived at Tooting at Mr Venables about ½ past nine o Clock. We had breakfast & I went to London.

23. Made an observation when I began making the Bank note paper. I have this week sent the whole of it off & it has given much satisfaction.

26. Went this day to Arundel (called on Mr Penfold, and agreed to take his Rags regularly).

27. Mrs S & myself at Storrington called on Mr Greenfield & Mrs Greenfield, (they keep different shops) about their Rags.

SEPTEMBER

27. [Tuesday] Dear James' [21st] birthday. We cleared out the Packing Soll and gave all the Men and boys a Hot supper of Mutton and Apple pies, some Cyder with their supper and a pint of Beer each afterwards. The Women & girls had tea in Pitfold finishing Soll. The whole number amounting to nearly 100.

29. Put in a new Hog wheel at the New Mill made by Harding the carpenter on the Lion Common.

OCTOBER

8. [Saturday] James went to Godalming for the Hog Brass.

17. The first time of coming to London after the reduction of the 1½[d] pr lb Duty off paper – I reduced about half – some houses were willing to buy others were not.

NOVEMBER

14. [Monday] James went to Rackham on his road to Arundel & Storrington about Rags.

17. James went to Portsmouth about Rags & purposed going to Fareham.

1837

JANUARY

12. [Thursday] Dined at my dear Mothers. In the year 1801 on this day my dear Father died.

> James's 85 year old mother Hannah Simmons *née* Philps was living in the High Street, Haslemere.

21. William began yesterday in the Glazing House – some of the boys not at work.

> James's second son William, who was 16 years old, was deaf.[8] An influenza epidemic was keeping employees away from work.

28. This week has been a very wet and heavy one – nothing dried in the least. We have a great many orders for paper waiting but cannot get it finished.

31. The weather just the same – no drying.

FEBRUARY

16. [Thursday] James went to Bignor & Rackham.

17. James return'd by Storrington, Pulborough, Wisborow Green & Kirdford – the Rags were all sold.

18. James went to Midhurst this afternoon about Rags.

MARCH

18. [Saturday] This week has been dull but no rain – yet heavy so that I could not dry any paper – my customers are very anxious for it – trade is brisk.

27. The Plumber putting in the pipe at the New Mill.

APRIL

29. [Saturday] The shaft of the Water wheel broke today. I had just got a new one ready – still hoping it would have gone till Witsuntide.

MAY

1. [Monday] Millwrights here about the shaft – cut the wall lower into the Wheel house and the steps partly down from the Engine House took the arms out of the first wheel and made the newel hole larger to get out the old shaft.

2. Very busy at the shaft. In the morning Puttick came up from Pitfold to tell me some rascals broke in the soll and stole some Rags – on looking more into it, we thought they took very few if any as they were not fines but London Seconds. They took the lead from Mr Appleton's part of the Mill, and what was against the water wheel of my part.

3. Got the shaft in its place and began hanging the water wheel.

6. This evening about 8 oClock finished in the Mill as to the new Shaft – it appeared to do well – & all done without any accidents.

15. Busy in the Mill putting a new beam the one nearest the Engine House – the tree came from the Hanger above the New Mill.

29. Abm Harding came up from the New Mill to say some person had broken in. I went down & found some of the lead torn from the Engine & Chest – the stuff made foul – the Finishing Pans & Hand Bowls stolen also the end of the Hog Pipe, the valves in the Cistern. Went to Pitfold they had also been there and stole the forcer – the water Cock to the Engine &c.

31. Went to Pepper Harrow, to the Revd Mr Elliott, obtained a search warrant for Deans House cart &c. After my return sent to Haslemere to call a meeting of the Society.

JUNE

1. [Thursday] Went to the meeting – suggested the necessity of sending for a Bow Street officer. It was agreed to send I wrote the next post.

4. Mr Goddard the Bow Street officer came and dined at Sickle Mill. He went off to Portsmouth.

5. Mr Goddard returned with lead from Mr Longs. In the afternoon we went to Hindhead – searched Wm Deans House – found nothing – and he declared he never was at Portsmouth in his life – consequently the lead was sold to Mr Long under a false name. We went to Winchester and looked over his house.

6. Had a meeting of the Committee. Mr Goddard bill of £7. 9s. 6d was paid.

9. Went to a sale at Chilworth Mills. Bought a few odd lots.

> Paper had been made at Chilworth since 1704. A new mill was opened there in 1836 and this was presumably a sale of equipment from the old mill.[25]

10. Busy this week throwing up the Mud in Pitfold Pond and making a single roof over the Engine House at Pitfold. Moved the Duster &c.

> The mill ponds had to be dredged periodically in order to maintain the maximum capacity for storing water. This was done in the summer when the water was low.

19. Mrs Simmons & Charlotte went to call on Mr C Lucas at Lowder Mill, married a short time since.

> The portrait of Mrs Simmons[8] reproduced as figure 7 was painted at about this time. James III usually refers to his wife in the diaries as 'Mrs Simmons' and here 'Charlotte' is his daughter. Lowder Mill (NGR SU900317) was a local corn mill.[9, 20] Its ponds and associated waterwheel symbol can be seen on the map of figure 1 on the County boundary south-west of Haslemere town. The Lucas family was related to Mrs Simmons.

JULY

6. [Thursday] Saw the Mr Longs on business, enquired of Mr Robt Long as to the lead he bought & also about his Rags.

12. We have continued at Sickle Mill to do our work as yet – trade is become more dull.

13. Rode over to Alton in the afternoon to see Mr Spicer about two Engines he has to sell. I offerred him 70£ – he wants 100 with the Hog & Pott.

21. It is now about 12 months since I sold the last of my land, and I have great cause to be thankful to the Lord. I have been since enabled to go on without so much anxiety in meeting my payments, and trade being better I hope I have been enabled to gain something.

22. Busy this week altering the Counting House and making a sorting Rags House of the place behind.

29. Finished the Counting House &c.

Figure 7. Portrait of Charlotte Simmons, wife of James Simmons III,
probably painted in the late 1830s when she was about 60 years old.
It is signed by E.S. Gibson who is otherwise unknown.
(Courtesy W.J.D. Cooper)

SEPTEMBER

23. [Saturday] The carpenters at work in the Drying House, making what was the finishing Sol a drying loft. The Millwright here also and moved the press from the sol to the Mill.

29. Returned from London by the Chichester Coach. Trade rather dull.

30. James went to Godalming to enquire about the Mechanical press.

OCTOBER

2. [Monday] Jas Carruthers the Millwright came over about the Mechanical press agreed with him for it for 90£ he to fix it up.

9. James gone to Alton to see Mr Spicer about his two Engines which he wishes to sell. I have offered him 90£ if he will bring them over.

18. Carpenters have been busy at Pitfold altering what was the finishing Soll into a Drying House. Ed Minor widening the water course from the wheel and cutting from the old waste water course to the corner of the Arch that comes from the waterwheel to fill the old course by the Rag House & Turf House.

24. Putting in an arch at Pitfold for the waste water.

31. James went with me to Barford to look at the Vatt Knotter which Robt Puttick put up. I think it does very well.

> William Warren of Bramshott Paper Mill is recorded as the new occupant of Barford Lower Mill on 27 October 1837.[5]

NOVEMBER

6. [Monday] I was busy last week in filling up the old waste water stream at Pitfold by the rag & turf House, and began on Friday to drain the piece above the Pond on the East side. We go seven feet [2.1m] deep and on to twelve feet [3.7m] and get a good deal of water, but not at present sufficient to supply the Engine for washing water.

17. James went today his round into Sussex about the Rags at Storrington Pulborough &c.

DECEMBER

1. [Saturday] This also has been an anxious week. Jno Harding leaves.

16. I went to Barford this afternoon to see how Robt Puttick got on with the Knotter.

23. Busy today taking out the second vatt press which I purpose to put up at Pitfold. Expecting the Millwrights to put up the mechanical press next week.

25. The Millwrights came to put up the mechanical press.

29. The new Press finished so far as Carruthers work goes. I am to give him 90£ for it fixed.

1838

JANUARY

12. [Friday] This week has been severe with frost & snow. The water wheels much frozen up and the half stuff so frozen that we cannot Bleach it.

FEBRUARY

3. [Saturday] Finished taking stock this week. I have prayed (some time ago) that the Lord would enable me to sell the Land, and that business might revive so that I might be enabled to continue in the same situation in the Mills as I had been. God has enabled me to do so to the present time & to repair the premises, and has also given me the prospect of doing more as I purpose putting in new tackle at the New Mill and setting on Pitfold ere long.

8. James went his round to Petworth Storrington &c.

10. We have a good many packs behind, & not able to size any paper for [blank] weeks. The New Mill quite still for the last fortnight, the half stuff being so frozen that we could not drain or bleach it.

17. On Thursday James hurt himself in breaking the back of the paper while jogging it & a little blood came from him.

23. It continues to thaw. The packs of wet paper begin to be a little mildewed at the top & bottom, we are now hanging some made in December – the two Engines &c. came from Mr Spicers, unloaded them at the New Mill.

MARCH

15. [Thursday] Putting in the slate Vatt at Pitfold. Robt Puttick also putting in the tackle for the Knotter & Hog.

24. Nearly put up the Knotter at Pitfold. I am often cast down by worldly thoughts & worldly troubles, sometimes I think if the Machines should so improve as to make better goods and at a cheaper rate than the hand made. What shall I do?

31. The Knotter finished at the Pitfold Mill – cost above 30£. I hope to set on there about Wednesday next. The small pox very much about. Jno Moorey who lived at Iping Head & worked for me so many years (tho' not lately) died with it.

APRIL

4. [Wednesday] Did not go to Church this evening – being very much engaged at Pitfold Mill – with the millwrights & carpenters &c.

12. Good Friday. The farming men did not work, the paper makers worked till it was time to get ready for Church.

18. Set on the Knotter at Pitfold to day, the pipe is not large enough to admit sufficient water to drive it with the Hog. We purpose tomorrow letting some water along in a shoot to help it for the present it appears to go very well.

20. I went to Barford Mill after tea.

28. Return'd from London by the Times Coach. Trade is very dull – no sale for paper & during my absence the screw of the Mechanical press broke & a fat calf jumped out of the stall and kill'd himself. These circumstances of loss are sent by the Almighty's direction.

MAY

1. [Tuesday] I cannot see my way clear in laying out the money at New Mill, I have a very considerable stock of paper by me, which if I had sold would have put me into cash to go on with it, but as I have not and trade is dull I told Jas Carruthers that I must put it off for the present.

JUNE

2. [Saturday] Fished Sickle Mill pond.

The pond with the mill buildings beyond is shown in the detail of the pen and wash sketch by John Wornham Penfold jnr., nephew of James III, reproduced as figure 8.[27]

4. Busy today sending out the fish, the Millwright & Masons also here.

16. I went up on Wednesday to Guildford and took the coach at half past six to Woking from which place the steam train runs from. I took my seat in the second Class, the ones next to the Engine and sat my back to it. I felt but little inconvenience from the smoke and ashes from the Engine and rode very comfortably, going up to Vauxhall in about one Hour and twenty minutes, what is from five to ten minutes longer than usual – the train began to carry passengers for about a fortnight. I paid 3/6 from Woking to London. I found trade dull in London – the coronation being about to take place seems to cause a flatness in the wholesale trade.

> The London & Southampton Railway had opened to the public from Nine Elms near Vauxhall to Woking Common on 21 May. It changed its name to The London & South Western Railway on 4 June 1839.[28]

18. Went to Barford about the Knotters.

JULY

4. [Wednesday] From Messrs Muggeridge &c. out of the last fine Copy [deleted at foot of page].

> Nathaniel Muggeridge followed by other members of his family were master paper-makers at Carshalton Paper Mill from 1817 until 1894 making hand-made paper with the 'C ANSELL' watermark.[29, 30]

27. Busy today turning the pipes & digging out the place for the water wheels for the Knotters to go in.

AUGUST

3. [Friday] This week we began putting up the Knotters at Sickle Mill.

11. Returned from London, trade dull yet I sold a few things.

17. Finished one knotter at Sickle Mill & purpose turning it on tomorrow morning.

OCTOBER

13. [Saturday] Water very short only done 9½ Days at Sickle Mill this fortnight at each Vatt. In the last week we have been quite still at Pitfold – Robt Puttick altering the Knotters. Had three men at the Pond throwing out the Mud and the carpenters & masons at Sickle Mill paving part of the Mill.

19. Busy at Pitfold altering the Knotter, laying a new leaden pipe in from the pond &c. which will not be finished till tomorrow – also lining the Iron furnace & new lining the coppers.

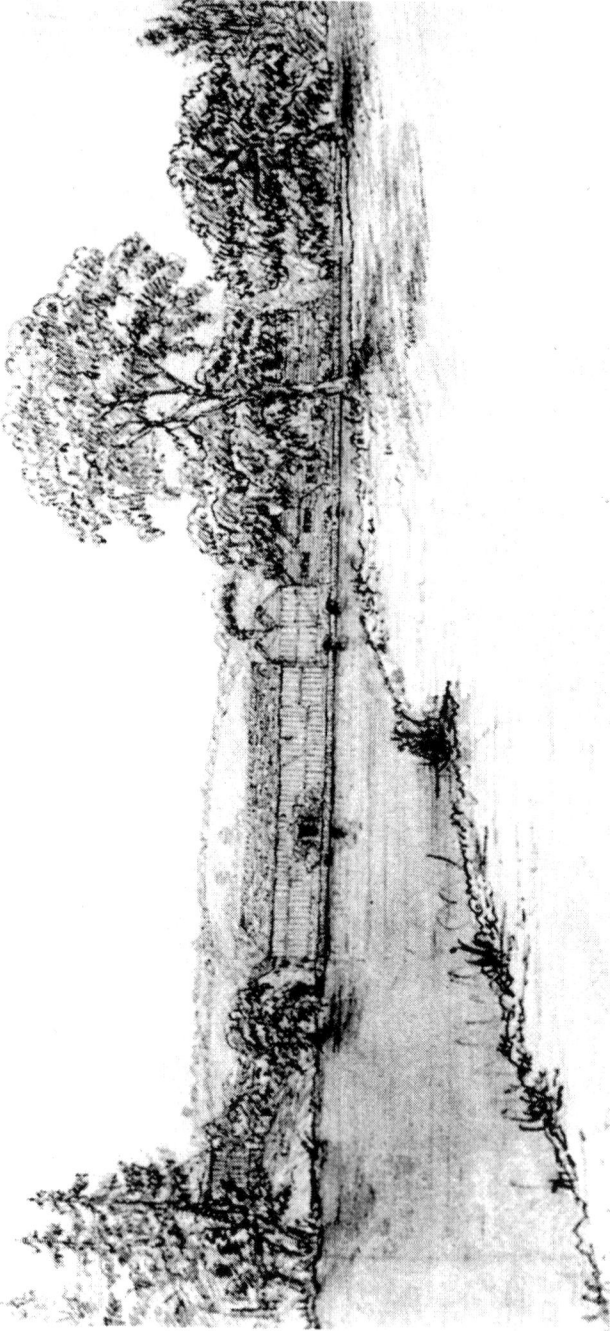

Figure 8. Sickle Mill from the north-east in August 1850.
The large mill pond, which has now been largely filled, appears in the foreground with the shuttered mill buildings beyond. Detail from a pen and wash sketch by John Wornham Penfold, jnr.
(Courtesy Haslemere Educational Museum)

27. The Masons & carpenters here new making the oven which was worn out – turned the mouth into the wash house. trade is very dull & I have a large stock of finished papers – 25 or 2600 Reams – yet I manage to get money to go on with & without owing much for Rags or other things – but do not sell sufficient to go on with the alteration at the other Mill.

30. The Millwrights came to repair the Glazing Machine – sent the rollers to Godalming.

NOVEMBER

5. [Monday] George went to Godalming for the glazing Rolls, where I had sent them to be fixed on the spindles.

9. Lord Mayors day – saw the procession along Bridge St from Mr Curtis counting House.

> The Lord Mayor's Show was held on 9 November from the reformation of the calendar in 1752 until 1959 when it was changed to the second Saturday in November to avoid traffic congestion.[31] Mr Curtis was a wholesale stationer.

12. Robt Puttick, putting up a turning lathe in the counting House.

24. The last few days dry – plenty of water now and sufficiency of orders having taken a large one of Mr Magnay with several other smaller ones coming in which will employ the 3 Vatts during the winter. I have now a large stock of finished papers on hand, which I hope will go off along the winter and if what I have to make answers it will I think enable me to do the repairs at the New Mill in the spring.

> William Magnay was a wholesale stationer in Upper Thames Street, became Lord Mayor of London in 1843-44, was created a baronet in 1844 and became insolvent in 1858.[32] With his brothers James and George he operated several paper mills including Postford and Stoke Mills near Guildford and Westbrook Mills at Godalming.[25]

26. Sent to Godalming for a slate Half Stuff Cistern for the Engine House at Sickle Mill.

DECEMBER

8. [Saturday] A fine day – put up one of Dr Arnotts stoves in the glazing room.

> Neil Arnott invented a smokeless grate which combined economy of fuel and consumption of the smoke with uniformity of combustion.[33]

15. The weather very fine and mild. But not drying much. I have a very large stock of paper which I do not sell – trade very dull – nearly 3000 Reams in storing, ready for market. If I had not taken a large order of Mr Magnay I should not have known what sorts to have made – the Lord is merciful to me. We are troubled in keeping the paper clean because the low Rags we are working have now a good deal of the makintosh stuff in it, with the India Rubber between, which is difficult to sort out and when beat with the stuff it goes through the Knotters & causes the paper to be full of black specks which will not pick out.

The shortage of rags had forced Simmons to use poor quality material including cast-off raincoats.

19. Went to Godalming on Monday for the Millwright – he did no come till last evening – & now he has not done the glazing Machine which was out of order.

1839

JANUARY

9. [Wednesday] Met James in London & went round the different stationers, trade dull.

MARCH

19. [Tuesday] Mr Daintry dined with us and afterwards bound Joseph Puttick and Charles Harding as out door apprentices.

As outdoor apprentices Puttick and Harding would continue to live at home. They would be bound to James III for 7 years.

23. I have a considerable stock of paper on hand, which I did expect would have gone off this spring at present it hangs heavy & no orders come in. I purposed doing before this the alteration at the New Mill, but trade being dull and nothing settled as to valuation of property under the new poor law act I have let it remain undone.

APRIL

1. [Monday] Mr Pewtress called and looked at our hand stuff & bought it, I took in part for exchange a Horse 9 years old for 30£ & a tun of Bleach 28£.

11. Trade dull although fine Rags are very much advanced, London Fines from 32s/ to 42s/ [per cwt] and no talk of raising paper. Dined with Mr Venables No 4 Arlington Street.

MAY

13. [Monday] After dinner Mr Allnot called at my mothers & had dinner. He is the son of Mr Allnot of [blank] Mill in Kent. He is now at Woking Mill which is about to be turned into a paper Mill and is to have part with Mr Aldm Venables in it.

> Henry Allnutt, snr. was at Lower Tovil and Ivy Mills near Maidstone in Kent.[12] Woking Mill appears to have opened in July 1840, the month in which William Venables died but Henry Allnutt, jnr. left 3 months later.[5] In 1855 he became the master paper-maker at Chilworth.[25]

20. Sent George for a slate cistern from Godalming – brought it but not done, oblg'd to send over for a man. Stopping up the door way out of the Engine House on the pond head and making a place for some Bleaching Chests.

30. Trade very dull many of the stationers about taking stock – could not sell.

JUNE

3. [Monday] This afternoon the screw of the Mechanical Press broke again.

11. A sad accident happened today. George Timms who had been in the glazing room but a day or two, he by some means got his hand under the rollers and it was drawn in and by that means so crushed as to cause it to be taken off. There was no paper under at the time, he had just done taken out the sheets from the coppers and the other boy was jogging the next parcel strait. I conclude he incautiously put his hand on the under roller and something took his attention another way while his hand continued on it and so was drawn in. I was in the Counting house with Penfold when the alarm was given and before I got up the poor boys arm was taken out. Mr Smith was soon down and return'd with Mr Gordon and performed the operation, the poor little fellow did not seem to suffer much.

> The diaries contain frequent references to accidents and illnesses which resulted in local medical practitioners being called. John Wornham Penfold, snr., was married to James III's sister Mary and was the grandson of James I.[6] Penfold's son John Wornham, jnr., was an architect and surveyor, and drew the view of Sickle Mill shown as figure 8.

26. Jas went to Alton yesterday to enquire about the Sizing. Mr Spicer not at home. Received a letter from him this morning to say Mess[rs] Hollingworth & Balston have raised their papers 10 pCent [?] and he should do so also.

JULY

6. [Saturday] It has been a week for great thankfulness to the Lord – likely to take a considerable order for paper.

9. Yesterday there was a Meeting of the paper makers & stationers to consider the plan imposed by the Government for the penny post establishment, as it was said that the envelopes were to be adopted and made by Mr Dickenson or that every sheet of letter paper was to be stamped either of which would be very hurtful to our trade. It was agreed to petition parliament to make the penny payable when the letter was put into the post and have no stamp or envelope. This caused trade to be very dull. I sold a few things and took some orders for more.

> The introduction of a uniform penny postage with Government stamped envelopes formed part of the Budget speech on 5 July 1839. A deputation of stationers and paper-makers went to see the Chancellor of the Exchequer claiming that they would be inconvenienced. In practice Dickinson was awarded the contract to make the paper for these envelopes as he had developed a method of embedding a silk thread in his paper which was an obstacle to forgery.[21]

18. This morning Elias Puttick and Isaac Salmon were not to be seen in the morning, and upon looking to their room it was found they had taken their things & gone off.

AUGUST

6. [Tuesday] Trade continues very dull. Money very scarce.

SEPTEMBER

4. [Wednesday] Took out the fish from the Vatt & Chest at the New Mill.

11. Mr & Mrs Venables came with two little girls & maid servants.

28. Bought a ladder & large Vatt at Jh Hardings sale of odd things at the corn Mill.

> Jeremiah Harding had been the tenant miller at Shotter Mill, which is marked on the map of figure 1, and was succeeded by George Oliver whose descendants remained there until 1939.[5] When Simmons offered the mill for sale in 1832 (see figure 6), it was said to have a powerful breast-shot waterwheel and two pairs of stones.[15]

OCTOBER

5. [Saturday] Trade dull – money market in a bad state – the penny post not yet arranged.

7. Mr Dunkin came in the afternoon about Jno Newman as an apprentice – he staid till the next day.

> Bryan Donkin was the engineer and inventor who developed the Fourdrinier paper-making machine at his works in Bermondsey. He also had an engineering works at Dartford.[34]

NOVEMBER

14. [Thursday] James went round his Sussex tour about Rags.

19. Went to London in the morning, business exceedingly dull. Set off to Dartford – drank tea at Mr Dunkin's.

21. Went into London early to see Mr Magnay.

29. Went to Godalming today about Pitfold Insurance and arranged it.

DECEMBER

5. [Thursday] Rode over to Sir C Taylor's yesterday to ask him about a summons for Charles Harding & Joseph Puttick – he recommended me to the Bench – called today at Colonel Webbs to get a summons, he was just gone out.

7. James went to Guildford today to enquire about a summons for Joseph Puttick & Charles Harding.

9. James went to Iping about Rags &c.

13. Told James & William they might have the old broken paper that laid about by paying me the duty when charged.

26. James went to Postford to see Mr Magnay about Rag dust.

> Postford Mill, on the Tillingbourne between Albury and Chilworth near Guildford, was acquired by the Magnay family in 1826 and rebuilt as the largest machine paper mill in Surrey. It was operated by James Magnay until he died following a fall from his horse in 1842 and then until 1852 by George and Jane Magnay.[25]

1840

JANUARY

7. [Tuesday] Set off about half past six oClock and drove to Woobourn arrived at Mr Lunnons about 3 oClock.

> Wooburn is in Buckinghamshire, four miles south-east of High Wycombe. Thomas Lunnon was the master paper-maker and miller at Hedsor and Fullers Mills.[5, 35] His daughter Ann married James IV on 16 October 1845.

8. Looked over the Mill, walked down to the other Mill – after dinner went to Burham – stopt to tea supper at Mr Jno Howard Junr and rather late home.

9. Walked up the opposite Mill. Afterwards called on Mr Fromow – met him going out to London by the railroad. Called on Mr Freeman Spicer & he was out, saw Mrs Spicer and she said there was no objection to my seeing the Mill – went from thence to Mr C Venables, he was out, & Mrs V unwell saw his son who said he could not show the Mill – drove round to Loudwater met Mr F Spicer, who wished me to come over the next day – then returned to dinner to meet Mr Howard & his son – after dinner we had some music.

> John Howard, snr., had been the tenant paper-maker at the Haslemere mills between the death of William Simmons in 1801 and the entry of James III into the business in 1811.[30]

10. The Mills on the Wooborn stand thick together. The Thames Mill is the first. Mrs Angels then

> (1853 Tom) Mr Lunnon - Hedsor
> Mr Peggs
> (Mr F E Venables) Mr Lunnon – Fullers
> (Mr Chas Venables Jnr) Mr Wright – (Soho Mill)
> (Messrs G & F E V.) Mr Ch Venables (Lower Glory Mills)
> Mr Spicer (Glory Mills)
> (Mr Geo Venables) Mr Frowmow (Clapton Mills)
> Mr Spicer – Hedge Mill
> Mr Gaveller
> Do
> Mr Plastow
> Mr Spicer
> Mr Morbey
> Messrs Lane & Edmonds
> Do
> Mr Edmonds
> Mr Lane
> Mr Fryer
> Mr Lane.

> Information in round brackets appears to have been added in pencil in 1853. Further information about these paper-makers and their mills is provided in the indexes.[5, 35]

11. This is the first day of the penny post – had five letters.

A uniform postage rate of 2d. had been instituted in London in 1801 but outside the Metropolitan area there were large variations. In 1837 Rowland Hill published a pamphlet advocating a flat rate of one penny postage for each half ounce for all letters posted and delivered in the United Kingdom.[21]

16. Trade very dull – disappointed in meeting with the people I wanted to see.

22. Went in the morning to Iping – Mr Pewtress not at home.

28. Trade is very dull, great distress in the manufacturing districts. The country very unsettled, Radicals, Chartists and Socialists trying to do evil.

30. Went to Farnham to attend the Magistrates against Joseph Puttick & Charles Harding. They would have been sent to prison, but I did not wish that, they were reprimanded and paid the expenses 11s/ on their promise to behave better for the future.

FEBRUARY

3. [Monday] Joseph Puttick' work not done well took out five sheet wrinkled & broken from about a quire and showed it to him – told him the fuel was wasting and his poshole dirty – he said he had no right to do all the poshole work.

'Poshole' appears to be a corruption of the French word pistolet used to describe a charcoal heater which was linked to the back of the vat and warmed the stuff. Looking after the pistolet was one of the dirty jobs done by apprentices. However as early as 1793 a clean steam heater had been patented.[2, 3]

8. Busy in taking stock and not done it yet. Trade very dull and things go heavy with me.

10. The Queen married to day to Prince Albert of Saxe Cobogh.

15. Busy yesterday & today taking out the press in the Soll at Sickle Mill which was broken and taking the one from the sizing house at the New [Mill] & putting in its place.

21. I have been busy moving the press from the sizing House New Mill to the Soll here – and taking down the old Vatt press at the New Mill. Sent to Godalming this day with the timber for the pitt wheel at the New Mill & brought back some of the iron work. I am at a stand what to do. The Vatt trade seems fast declining, I cannot sell the paper I now make, having one Vatt now still – after the New Mill is done unless the trade alters it will be of little use. I have a Desire to put in a Machine but the expence is so great that I cannot accomplish it.

An illustration of a paper-making machine dating from this period is presented as figure 9.[36] In 1840 there were 462 paper mills in the United Kingdom with 191 machines and 380 vats. A year earlier there had been only 129 machines.[12]

28. Went to Godalming to Carruthers about sending up the timber for the water wheel & settled accts with Mr Stedman.

MARCH

7. [Saturday] We began to hang paper but were fearful the frost may touch it.

10. Went to London by the railway. Trade dull.

12. Return'd home having done little business.

14. I wrote to Absolam Atkins to say as James was now wanting to get more into the details of the business I wished him to take his place & that he had better look out for another situation. May the Lord direct & guide us all – trade is very bad.

Atkins appears to have been a senior clerk. James IV was now 24 years old, only four years younger than James III when he took over full responsibility for the business from John Howard, snr., the tenant paper-maker at the mills from 1801 to 1811.

19. Borrowed a spirit level of Mr Stedman to take the fall from the tail of the New Mill to the end of the New Moor piece and found it to be more than 4 feet [1.2m] which is two more than we calculated on. I made up my mind last autumn to make the alteration at the New Mill, hoping this spring that I should have got off my stock of paper – but I fear it; as trade is very dull & at present no sale.

Again Simmons is planning to increase the head of water at New Mill. When the mill was for sale in 1854[8] there was a powerful breast-shot waterwheel 16 feet (4.9m) in diameter and 9 feet (2.7m) wide and the head of water was said to be 9 feet (2.7m), which was 4 feet (1.2m) more than in 1832.[15]

21. Trade continues exceedingly dull – no sale for paper, which inconveniences me very much.

23. Met the Revd Henry Baker this morning in the moor piece to talk with him about cleaning out the stream below, he said I might cut it strait and dig it out as I pleased, so as not to put him to any expence.

27. Pulling out the old tackle at the New Mill, set out where the stream is to be new cut. Trade being so dull I cannot sell paper sufficient to get money to pay as I want it. Yet I still hold on & get credit. If I could sell my stock of finished paper I should have sufficient and some over.

28. Fish'd New Mill Pond & took out the fish, put some into Pitfold, some in the stream (cut) and some in Sickle Mill Pond. Began cleaning out the stream below.

APRIL

1. [Wednesday] Busy digging out the stream New Mill.

2. James gone to Arundel and round to Storrington and Villages.

4. Gave three men 30/ for diggin out 30 yards [27m] the lower piece about 200 square yards [167m²].

7. We get on with the stream, but some work to do before it is finished.

13. Absolam Atkins went off today, to his new situation with Mr Curtis.

14. Digging out the stream at the New Mill the upper part 4 feet [1.2m] deep, 10 feet [3.0m] wide at the bottom, all gravel as yet and very hard.

16. The work at the New Mill occupies much of my attention.

18. The weather just suitable for our work at the New Mill. It is more to do than I thought, as we gain 4 feet [1.2m] fall instead of two [0.6m], which is of great advantage if the trade will answer anything.

21. Mr Daintrey dined with us after the turnpike meeting. I did not go up as I was busy at the stream New Mill.

24. James gone to Winchester to a meeting of the paper makers about the price of Rags. I am still very busy at the New Mill. We get on with the stream and I hope to finish it next week. It will cost me more than I thought yet I hope it will be a pretty good job – not much less than 20£. [later insert] (I found it afterwards very much more).

29. Trade exceedingly dull – no sale for paper.

MAY

8. [Friday] Trade still dull, very little sale.

9. Received yesterday morning two of the new envelopes with letter enclosed. The devices are rediculous.

The Mulready envelope, with an elaborate engraved design featuring Britannia and a host of other figures including two elephants, went on sale on 16 May 1840 together with the penny black and twopenny blue stamps. It was rejected by the public and nearly all of the vast number produced had to be destroyed after six months.[21]

11. Today Jas Carruthers came and set out the fall and work to be done at the New Mill – digging so deep and some water at the lower part causes the earth to give way and we have been all day with 5 men getting it out. It is an anxious time. I hope tomorrow to begin getting in the foundation. Trade is very bad – no sale for paper & no orders except one just now for Ionian F-Cap for Mr Magnay.

> Simmons used some of his Ionian paper for diary 23 in 1842. A photograph from a page of this diary taken with transmitted light is shown in figure 11. It reveals the laid and chain lines of the mould and part of the watermark consisting of the winged Venetian lion of St Mark surrounded by a garter bearing the words 'United States of the Ionian Islands' and surmounted by a crown. The Islands (Corfu, Paxos, Leukas, Ithaca, Kephallenia, Zakynthos and Kythera) were captured by the French from Venice in 1797 and then following an unsettled period they became a British protectorate in 1815 before being ceded to Greece in 1864.[37] Ionian Islands postage stamps and coins are well known to collectors as is the fact that cricket is played on Corfu.

14. Showers but not to hurt us much at the New Mill. Got the foundation in & wall up four feet [1.2m] on the south side. I trust out of danger from any slips of earth.

15. Got the wall up for the pillow block – propped the side of the Mill up, moved the Engine, took out the wall, settled to underpin the old wall and not take it out.

16. Dug out the place for the pitt wheel.

22. It has been and still is a very anxious time with me as regards temporal concerns. I am going on with the repairs at the New Mill, the expence heavy and no trade in paper, the sale never more dull than at the present time, so that I cannot get on as to money to pay as I could wish. The weather has been favourable for what we had to do at The Mill and no accident has yet happened.

25. Digging down for the arch over the thor's [?] below the water wheel. A Duty of 5pCent is laid on all excisable goods, consequently on paper – & that has taken place immediately.

30. Finished the arch at the New Mill.

JUNE

3. [Wednesday] Yesterday – Went to Bermondsey to look at the Fountain Paper Mills, which Mr Turner lately work'd by steam – it did not answer and all the Machinery & Plant is to be sold by auction next week.

George William Turner had been at Fountains Mill in Cottage Row, Bermondsey since 1829. He was bankrupt in 1835 but was able to continue at the mill until February 1839 when he was bankrupt again.[5]

12. Getting the wall up for the timber to put in there for the Engines &c.

23. I have not had an order for paper for a long time.

26. Mr Spicer from Alton came over about the price of Rags; agreed to lower them.

JULY

2. [Thursday] Got the rolls from the pond to their places in the Engine House.

11. Mr Spicer came over about the meeting at Winchester.

17. James went to Winchester today to a paper makers meeting to consider the price of Rags Wages &c.

18. James return'd after dinner the men waiting to be paid.

AUGUST

1. [Saturday] Mr Wm Hewitt came. He told me he saw in the papers that Mr Alderman Venables was dead – he was at the Isle of Wight.

24. Still at a loss what to do in getting a machine. Mr Lunnon recommends one of Messrs Donkin & Wilks they are much better than the Scotch or any other in his opinion and I think it may be so – the sum required would be at least 1300£.

The machine at Fountains Mill, Bermondsey must have been built by Bryan Donkin at his nearby engineering works. In Scotland paper-making machines were being built successfully by George and William Bertram and by the 1860s their firm and that of their brother James were sending many machines to England.[26]

SEPTEMBER

5. [Saturday] Trade still at a stand – no sale at all for paper and we scarcely know how to keep on.

16. Went to London to see about the Machine at Bermondsey. Look'd at it and agreed to meet Mr Turner. Trade very bad.

19. James went to Midhurst to see Mr Chorley about moving & putting up the Machine.

22. James went off for London with Mr Chorley to look at the Machine.

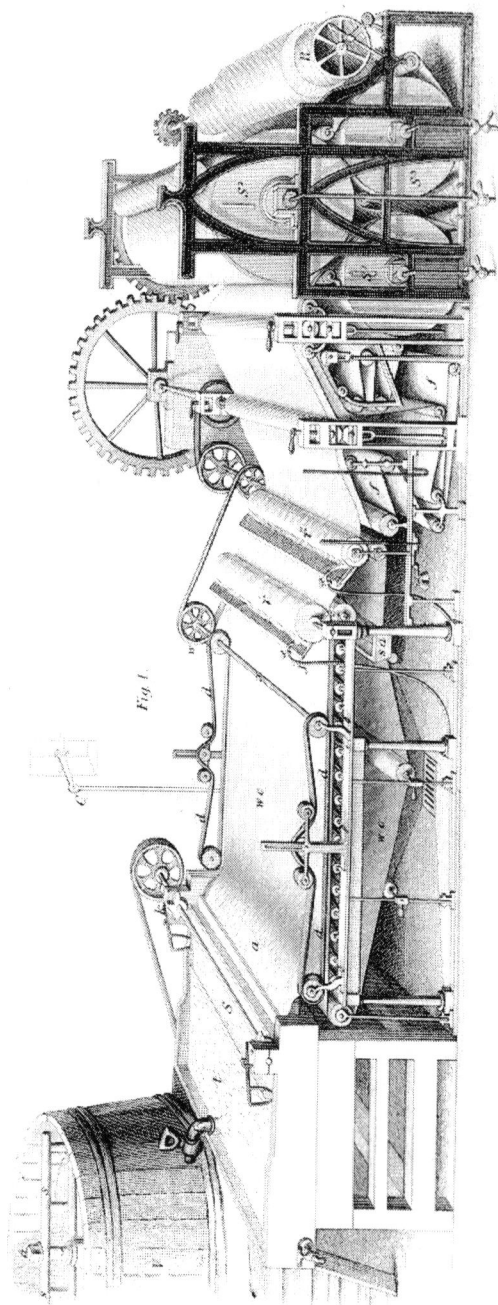

Figure 9. Fourdrinier paper-making machine of about 1840.
The stuff flowed from the timber chest at the left through a spout into the rectangular vat and then on to the endless web of wire shown at the centre. This carries the wet paper to the felts, rollers and drying cylinders at the right. (Steel engraving from Tomlinson's *Cyclopaedia of Useful Arts and Manufactures*, Vol. II, 1852 page 365.[36])

24. Went of to London to meet Mr Turner about buying the Machine, we could not agree for some time Mr T wanting 400£ – I offered him 375£ – to be paid for by Bills at 6 & 12 months. The Machine was to be complete in all parts & delivered up to me on the 5th Oct after which they gave up all responsibility. I calculate it will cost me as much more to get it up with all the appurtenances of steam boiler, water wheel, stuff chest (there is a stuff chest but it is old) &c. I hope I have done right in buying it. I had when I took stock this year 3000£ after paying all that I owed; (except money I owed on interest) trade since has been very bad & I have no doubt lost money, besides paying the repairs of the New Mill – so that I calculate I shall not have left in my paper trade (besides the Farming Stock and paper utensils such as felts mould &c.) 1000£ which will not be sufficient and to borrow money goes very much against my inclination.

26. Mr Chorley came up to set out the ground plan of the premises for the machine. He purposes going to London on Monday to take it down, and pack it, we are send our waggon up on Tuesday and see Moon about sending his on Wednesday.

28. Gave notice to the Men on Saturday that I should lay the Vatts still at the end of the fortnight for putting up the Machine. A trying time for both Master & men. Robt Salmon came to see me this afternoon for a character as Revd Mr Carey wanted a man to look after his Horse and work in his garden. Robt has been with me now for about 30 years and a faithful servant.

30. Nearly done the stream below the New Mill. I think we shall get it sufficiently low to carry off the water very well.

OCTOBER

6. [Tuesday] Went off to London early. Went to Bermondsey to see about the Machine. Mr Chorley & his men packing – staid and helped load our team. George being up with 2 Horses.

7. Went into town in the morning and then to Bermondsey. Moon waggon went down to take a load of it.

9. Dined at Mr Daintreys – talked with him about some money I wanted to raise to pay for the Machine &c.

10. Looking over my Stock Book this morning, I found a mistake I made at the time of taking it in Feby last of very considerable amount so as to place me in a trying situation for if I lost so much more money in my trade last year, than I thought I had, how much more must I be deficient this, having only two Vatts at work and the trade very bad.

13. Busy yesterday and today taking out the Vatts, Chests &c. in the Mill – expecting Mr Chorley to set out the place for putting up the Machine tomorrow.

15. Began taking away the walls & moving the earth for extending the building for the machine.

16. When I came home the Team was here with the cylinders &c. of the Machine. We got them into the Mill, without any accident but not till after dark.

17. Mr Spicer came over. He is to let me know about a boiler. Miss S Appleton married to Mr Pewtress Newington Causeway.

19. A fine day for us in getting up the building for the Machine.

22. Put on the roof to the new building for the Machine.

24. Mr Spicer came over to tell me about the price of a boiler – one of 9 Horse power 55£.

27. Mr Chorley came over to set out the foundations for the machine. He is to make the water wheel Iron – 15 feet [4.6m] high 2 ft [0.6m] wide – with shaft & spindle to drive the machine – the mouth of the Iron pipe for 70£.

> When Sickle Mill was for sale in 1854 the particulars stated that there were two waterwheels, a powerful overshot wheel 17 feet (5.2m) in diameter and 7 feet (2.1m) wide and an iron wheel 15 feet (4.6m) in diameter and 3 feet (0.9m) wide.[8] The latter is clearly the one being proposed by Chorley.

28. James returned from London quite well – sold some paper though at a low price. I sent Teasdale over to Mr Spicers to look at his boiler chimney to make ours by. Mr S writes me word that he has ordered a boiler for me but the price is 65£ and not 55£ as he understood before it is to be 8-6 [Probably 8 feet 6 inches, 2.6m] long of the best Iron and stout.

29. We got on with the work for the Machine house.

31. Finished getting out the earth so as to leave room for a person to get under the machine and began making the side walls for it to bear on. James went to Midhurst to see Mr Chorley about the water wheel and about sending his men on Wednesday. I have sold more paper this week than I could expect though some at a low price & since James returned from London we have had several orders.

NOVEMBER

1. [Sunday] Robt Salmon has left us for the last week or two, as he was out of place when we laid still he is gone to Iping to Mr Pewtress.

11. Mr Chorley came yesterday and set on two men putting up the Machine.

13. Very heavy rain. Dug out the foundation for the Steam boiler Chimney yesterday afternoon, but could not do anything today – but get down to a large stone 5 feet [1.5m] over each way which we purpose for the chimney to stand on if we can get it in.

17. Heard yesterday of extraordinary high water at Portsmouth, Southampton &c. Mr Lucas at his Mill at Bedhampton has been damaged it is said to the amount of 200 or 300£. Could not get the stone in for the foundation of the Chimney, but got it over the hole ready to let down.

20. The day fine got on with the Chimney – laid down the large stone on which it stands and put up the scaffold poles &c.

25. The last day or two very fine and mild, getting on with the Boiler Chimney.

27. Took down the Mechanical Press to make room to set out the place for the Stuff Chest. The masons did not work yesterday for the Chimney to get a little settled – at work today.

28. Busy taking down the Mechanical press. James went to Mr Spicers. The boiler to be done on Monday.

DECEMBER

1. [Tuesday] Robt Puttick has been here today taking down the old press in the Mill and preparing for getting the other two above.

2. At the presses all day.

3. Still at the presses, got them up nearly to the height we want them.

4. Got up the presses in the morning and removed the Mechanical press to the New Mill.

9. The Boiler came this evening & we unloaded it by Moonlight. Placed all the drying Cylinders in their places today.

10. Got the Boiler into its place.

12. Mr Chorley came to set out the position of the water wheel &c.

17. Returned this evening from London – came down by Chilman as I feared the rail train would be hardly able to travel on acct of the snow. I went to search for the patent paper cutting machine to read the specifications. I was told to go to the patent office Lincolns Inn. I there found the patent was granted to Mr Cowper in the year 1828 and that Mr Magnay had purchased it. I then returned to the enrollment office and read it – it speaks of the patent as applied to the cross cutting knife with a saw edge, but describes the whole machine but do not claim it all as a new invention. I went about the pipes also for the water to the water wheel and some small pipes to bring it from the Garden to the Boiler &c.

> Edward Cowper's paper cutting machine, as installed at Magnay's Postford Mill, was described as an extremely beautiful contrivance, its novelty being that it cut paper to a uniform size.[25, 38]

1841

JANUARY

18. [Wednesday] Sent George for the Iron wheels for the machine – hear today the gudgeon of the water wheel broken on Saturday evening.

23. George Newman is wishing to take a windmill near Chichester. I wished to dissuade him from it. After I left today Mr Spicer came over and staid the night.

FEBRUARY

4. [Thursday] After breakfast set off for Portsmouth called on the Longs about rags and at Mr Comerfords about paper.

6. The frost continues and the snow drifts – it is very severe, We can do nothing with bricks & mortar.

12. Went to Godalming and looked at the machine at Catteshall Mill.

> After being bankrupt in 1830 Thomas Sweetapple became the tenant papermaker at Catteshall Mill, Godalming. He had a Donkin paper-making machine installed in about 1837 and patented an improvement to this machine in December 1838.[16]

16. We have at last got most of the things for the machine – we might have got on faster if Mr Chorley had put on more men.

17. Went to London. I found rather more sale for paper.

18. Sold rather more paper than I expected.

22. The weather mild & pleasant – making some progress with the machine.

23. James & William went to Alton – William with the intention of staying a few days. But Mr Spicer being out they returned home.

26. James gone to Arundel for to see about some coal.

Arundel was a significant port and coal unloaded there could be taken by barge along the Arun and Rother Navigations to Midhurst, only 7 miles south of Sickle Mill.[39]

MARCH

5. [Friday] Yesterday morning, while driving the two Engines at the New Mill (the upper Engine the first time with wrapper stuff) ran out of gear and done a great deal of mischief – 70 cogs broken in the pitt wheel & abt 20 in the fly wheel.

These cogs would have been made of hard timber such as apple, pear or beech.

6. Mr G Warren called and William returned safe from Mr Spicers.

13. I did expect to set on the machine on Monday morning, but it is not quite yet done. It is very trying to be kept so long still.

22. Between two & three oClock the Bishops Carriage drove up. He had lunch then went in to the Mill and saw the Machine and said he must come again and bring Mrs Sumner to see it. This day we are making some wrappers with the Machine & getting on pretty well – we began last Tuesday 16 March a little & gave the men some supper.

Charles Richard Sumner was Bishop of Winchester from 1827 to 1860.[32] He was normally resident at Farnham Castle where Simmons visited him in connection with the building of the church at Shottermill.

27. We have been making small Hand with the Machine & get on pretty well. Mr Chorley came over to see if the Chains can be done away with and the wheels of the cylinder removed to the other side.

APRIL

2. [Friday] We have been getting on very well making Lunberhand the last two days and today making a few more wrappers and doing it well. We have also one Rag boiler in and hope soon to be getting in orders.

6. Set on this evening with the Machine for demy. It seems to do very well.

7. Mr Simmonds from Borne Mill came to look at our Boiler and Machine.

15. Mr Chorley put up the wheels to alter the going of the Cylinders.

16. Jas Carruthers came today to set the Engine to rights.

22. Last Thursday Charles Andrews and an apprentice came over to put up the riggers &c. for the alteration in the machine and it is now past 8 oClock and they have not done.

30. Came home by the train. Could not do business had some of my Machine Demy up – but no buyer. I am stretched for money. I did hope to sell and scrabble on; but I could not.

MAY

3. [Monday] Elias Puttick out of his time today and gave him his indentures.

5. We do not get on with our machine, what with little difficulties, it hinders us and the paper wrinkles.

6. The machine went on pretty well today.

7. The machine went very badly today – left off and sent for Mr Chorley.

8. The machine after the morning went very well. Mr Chorley came up. He thought it might be in the press rollers.

15. James gone to Wrackham and around to Storrington about Rags. Trade is exceedingly dull both for Vatt & Machine paper. I have about 2000 Reams finished which if I could sell would enable me to pay my way, and then if I could get order for my machine paper (by the blessing of the Almighty) I should get some returns of the expences I have been at.

20. James sold some paper but at a low price.

26. Busy at the New Mill, putting a bottom strainer in the Engine, fixing the lead pipe &c. and put up shutters over the gear work.

JUNE

10. [Thursday] Yesterday and today we have been grinding the press and glazing Rolls of the Machine down with emery & oil. Robt Puttick went home and from not putting on sufficient oil while he was gone, the emery cut into the Iron so as to mark the sheet, that we are oblgd to turn them to get it out.

15. Set on with the machine this afternoon only, it taking so much time to do the rollers. I hope we shall now go without any hindrance, it appears to do very well.

16. James went yesterday and today to the sale at Chilworth Mills all the plant and machinery to be pulled down and sold. James bought two old Engines for 14/- also some windows for the old loft over the Mill.

The new mill opened at Chilworth in 1836 (see 9 June 1837) had closed by 1838 and this was presumably the resulting sale. The mill reopened in 1846 and continued until 1870.[25, 40]

JULY

1. [Thursday] Trade exceedingly bad.

3. The Machine does not go on comfortably. The straps are old & break – the felt also is troublesome. Wants a New one but it is not come.

7. Wrote to Mr Tyrrell to say I would take Thos Lunnon at 100 pr annum.

Thomas Lunnon, snr., of Hedsor and Fullers Mill, Wooburn, Buckinghamshire, had died in March 1841 and his son Thomas was coming to stay at Sickle Mill to learn the trade. His representative could have been Edward Tyrrell, Remembrancer of the City of London from 1839 to 1869. Thomas Lunnon, jnr., later became the master paper-maker at Hedsor.[5]

10. Put on a new felt and a new wire to the machine and also a jacket to the press roller.

13. James drove to Farnham about the Rags.

14. The Machine at times still troubles us – we do not get on well & I fear not profitably with it.

30. James return'd in the evening – trade very bad. Borrowed a pair of press Rollers of Mr Warren and Mr Chorley came about putting them up. We do not get sufficient surface to our paper on both sides.

AUGUST

4. [Wednesday] James went off for Wrackham Storrington & from thence to Chichester and Portsmouth for Rags.

5. In the afternoon began making a News Demy for Goovenn & Co. It was some trouble to alter the cutting machine for so large a sheet, and in taking out the long screw to do it, it was broke – did not begin till 3 oClock.

12. Sent down Jno Croft in the morning to the New Mill with the mare & cart with Rags & to bring back half stuff – at Breakfast time Boxall came up to say that the mare as she was coming by the corner of the drying house ran off and not taking sufficient room over the arch way on the pond head the cart ran over side of the arch & pulled the cart & mare into the floush hole below. The mare was got out without hurt – the cart had one thill broken. The stuff we got out but a good deal wasted. When I got home to dinner I measured the paper then making, too large and not clean. The lead pipe off at the Rags boiler.

16. I went to Mr Warrens and drank tea.

17. We have been very much perplexed with India Rubber and rubbish in our paper, which has distressed me as we want it in the market to get the cash to go on with. I have a great deal to pay (for me) next month and things go very contrary.

26. Came down by the ½ past 5 oClock train – trade very dull.

SEPTEMBER

18. [Saturday] James went on Thursday to Farnham to get a summons for Charles Harding, for not attending to his work & refusing to do the Rag boiling.

22. Mr & Mrs Pewtress came just before dinner, and dined with us and then went on to Iping.

27. The Millwrights [came] to put in the rollers – the rollers we had from Mr Warren (he lent them to us) – our surface was not good enough without them.

29. James rode over to Bramshot to Mr Warrens & Mr Chalcrafts.

30. Went to Farnham having summoned Charles Harding to the bench for not doing his work and saying it was not his place to boil rags and go with the cart. The Magistrates settled that it was.

OCTOBER

1. [Friday] Finished this evening putting in the press rollers.

2. Exceedingly vexed when I got up this morning to find the press rolls were not done at least some little jobs about the Machine as I was very anxious to get on as we had a little order for some more News Demy.

9. This week have been very wet and we have done more work at the Machine.

12. Today getting on at the Machine with plenty of stuff about 11 oClock the wet felt from some cause got so torn that we have not been able to get on & now 7 oClock at night. It is very trying. James went off for London early this morning, purposing (DV) to go to the Sale at Fullers Mill.

> This was a sale following the death of Thomas Lunnon, snr. A photograph of the two facing pages of diary 22, showing the full entry for this day together with those for 15 and 16 October is reproduced as figure 10. A full transcript is given in the caption.

16. Made 100 Rm Demy at Machine yesterday & today.

18. The wet felt being old we took it off and put on another which had been on before but was mended, it took us half a day before we began work and then could not make it do, and was oblged to shut down. We hope to get on the new felt James bought at Mr Lunnons sale tomorrow morning in good time. In my situation these things try me very much.

> Simmons is clearly suffering from the problems inherent in buying second-hand equipment.

20. James went to Alton to see Mr Spicer about some Rags.

21. Recd a letter from Mr Spicer to say Thos Lunnon would not come to us on Saturday as we expected but on Saturday after he had some papers & accts to look over. Some furniture for his room and a dog came today.

> This refers to James Freeman Gage Spicer who was operating Glory Mill, Wooburn. Shortly after the death of Thomas Lunnon, snr., this mill was purchased by his daughter Emma and her husband Edward Fox.[5]

30. The weather continues very wet – we have more water than we use, just now. Trade very dull. James went to Guildford to meet Thos Lunnon – he is come to live with us. I pray God that it may be in harmony & love – that we may be enabled to guide him to the Savior as his only refuge while we teach him the trade he is to follow.

NOVEMBER

2. [Tuesday] Trade very dull – a disastrous fire at the Tower.

13. Wm with Thos Lunnon went over to Mr Warren's about a load of sacking.

15. Mr Warren came over about some paper – his brother Andrew with him.

22. Our Machine lately has gone very well but today a strap broke and again just as they were shutting down one of the Rollers of the dry felt broke asunder in the middle and tore the felt nearly all across.

27. Could not get on with the Machine – in the evening just before dusk we began but the paper wrinkled. Turned the spring water at the New Mill into the tiles to bring it higher into the cistern but they would not carry it.

Monday Oct 15th

Tuesday Oct 1st

'Tuesday Oct 12
Very wet. To day, getting on at the Machine with plenty of stuff about 11 oClock the wet felts from some cause got thin [?] and so torn that we have not be able yet to get on & now 7 oClock at night. It is very trying, but under little as well as great afflictions we must see the hand of the Lord in it & submit patiently to his will using the lawful means & perhaps this I have not done. James went off for London early this morning, purposing (DV) to go to the sale at Fullers Mill before he returns.'

'Friday Oct 15th.
A tolerably fine day after a wet night. My sister Mrs. A Newman came up to Haslemere yesterday to see her mother. White drove her – and slept at Sickle Mill. I went up after tea; dear mother was poorly when I first went in but afterwards was better. White Newman and Ann rode over this morning to Woolmer Lodge to see the preparation for the dinner to be given to day to the poor of the Parish by Sir Archibald Macdonald who is of age; after their return White went to Haslemere to dinner and drove his mother home We did not see her at Sickle Mill. Saturday – Oct 16 – Wet day – Jas. returned at tea time – well – Made 100 Rm $^{14}/_{15}$ Demy at Machine yesterday & today.'

61

DECEMBER

1. [Wednesday] The times are very distressing. A week or ten days ago the Chichester Bank Ridge & Co stopped. Since then the Petersfield and the Arundel Bank. Trade very dull with me, so that I can scarcely sell any paper to keep on. Today I had a letter to say that there was a misunderstanding – about some paper James sold & they could not take it – this disappointed me!!

8. Mr Savage & his brother came and bought some felts and moulds &c. I am pleased and thankful to the Lord for it. It was dead stock & I want money very much.

> Now that Simmons had ceased making hand-made paper he had no need for his moulds and the associated felts.

15. Went to London. Business very bad.

18. Very unhappy about the state of my affairs. Made a rough calculation of debts &c. and found that I have no more than will pay all I owe.

30. I went to Alton and dined with Mr Spicer.

1842

JANUARY

15. [Saturday] Isaac Salmon, has lately done his work so badly, that I paid him for the eight days till his time was out and gave him his indentures & notice not to come on my premises again, if he did I would prosecute him.

> Salmon who was born in 1820 was apprenticed on 30 June 1835 and had therefore served less than seven years.

28. Got through my business (trade very dull – could not sell any paper or get any orders) and came to Vauxhall to dinner.

FEBRUARY

12. [Saturday] The day wet – busy at the New Mill, putting the Engine &c. to rights – the lead wears very fast from this press [?].

19. Trade still very depressed.

MARCH

2. [Wednesday] Mr G Warren came over to buy Rags.

16. Business dull. Came to Vauxhall in time for dinner.

APRIL

7. [Thursday] Trade very dull can scarcely sell any paper. Have now none sold & no orders.

13. The new Tariff appears to make our trade still worse the import duties being lower it is considered detrimental inasmuch as some may be it is thought imported from France to this country.

> The Import Duty on first class paper had been reduced from 9d. to 4½d. per lb., plus 5% in each case (see 25 May 1840). The duty on imported rags was also reduced from 5s. to 6d. per ton.[12]

MAY

7. [Saturday] The Knotter has troubled us at times for the last week past. Trade very dull, nothing doing did not sell any paper in London.

30. Trade is exceedingly bad. I find it very trying at this time scarcely knowing what to do for the best. This I fear, that me & part of my family do not serve and honor God as we are bound to do.

JUNE

4. [Saturday] Busy yesterday & today weighing Rags &c. – to look over my stock again.

8. Return'd in the Evening from London trade still in the same state.

18. Every week and every day is a most anxious time with me now – trade being so bad that I can scarcely keep on and loseing money by it – so that if it does not alter I cannot tell what I shall do.

23. James went yesterday to Plester [Plaistow] to see for some Rags.

30. James returned safely from London – still no sale for paper, got a small order or two with the hope of a larger one.

JULY

7. [Thursday] Mr Warren and his son came over to dinner & buy our coarse Hands & Rope. They told us that Mr Spicer of Alton was married again today to a Miss Wright.

> John Edward Spicer, jnr., had seven children by his first wife Charlotte, *née* Pigot, and one by his second wife Sarah, *née* Wright.[22]

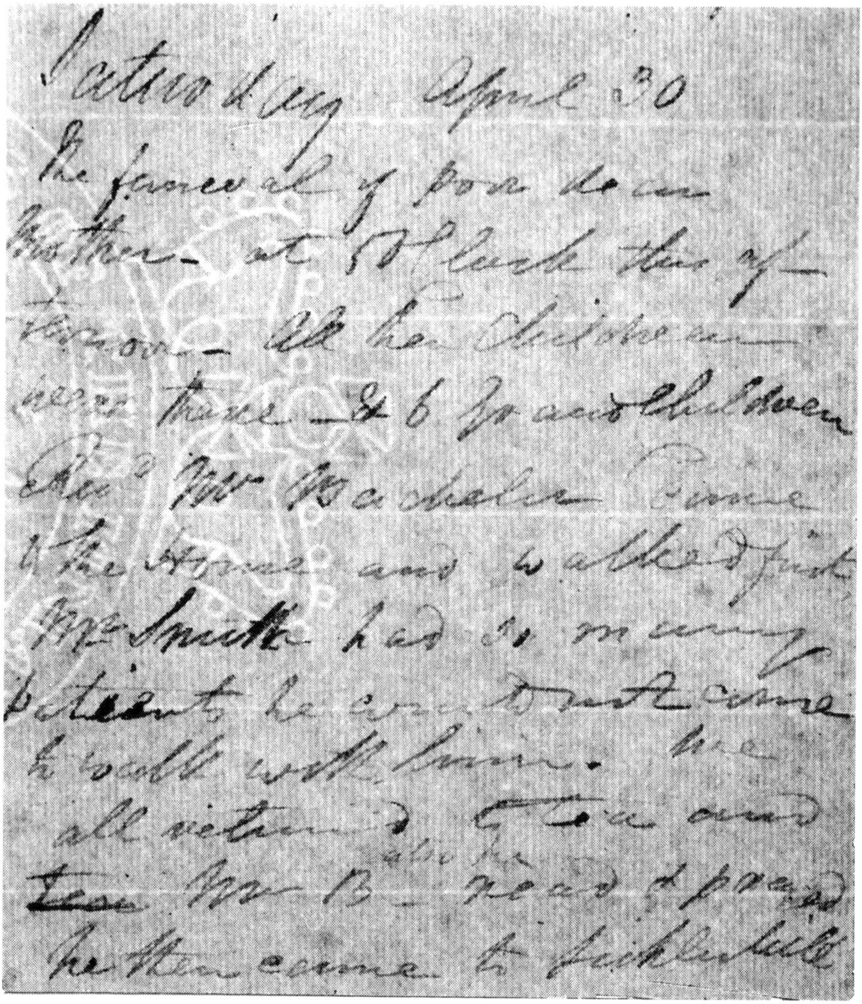

Figure 11. Photograph taken using transmitted light of a page from diary 23, covering part of 1842. The paper is hand-made laid, and displays part of the Ionian Lion watermark which is illustrated fully in figure 26. Simmons mentions this paper in his diary on 11 May 1840.
(Photograph by Gareth Crocker)

Transcript: 'Saturday April 30 The funeral of poor dear Mother – at 5 oClock this afternoon. All the Children were there & 6 Grandchildren Rev Mr Bachelor came to the House and walked first. Mr Smith had so many patients he could not come to walk with him. We all return'd to tea and Mr B. read & prayed he then came to Sickle Mill.'

14. Set on the upper engine at the New Mill with the new Washing Cylinder in it – it does very well.

20. Could not get out of London last night being detained by Magnays settling the acct.

AUGUST

12. [Friday] Mrs Ewen with her Father & Sister came to see the Machine.

13. Sent William to Godalming for some money to pay the men. The bank had not advice of some money that was paid in and did not send it which distressed me much.

18. Return'd from London – trade exceedingly- dull and what I sold was at a ruinous price.

21. We have had the new Knotter in at the machine two or three days it is some trouble to us – one of the corner pieces broke, it splashed very much and do not let the stuff through freely.

22. The Knotter does rather better. We altered it this evening making more room on the back side.

24. The Knotter now works very well.

26. I am still hampered in my concerns. I have yet been enabled to honor my Bills & pay my wages, though with much difficulty – the Bank not showing me any favor.

SEPTEMBER

1. [Thursday] Put on a new wet felt today.

24. Had just sufficient money to pay the mens wages.

OCTOBER

6. [Thursday] Josh Puttick had a severe blow in the neck from the falling of a rack full of wet stuff.

11. This morning the boiler leaked from the water being low overnight so as to heat the flue which caused the stone [?] to fly to pieces that regulates the water that comes into it.

NOVEMBER

7. [Monday] James went off to London this morning. May the Lord prosper his journey, which of late has been very sad; trade being in a deplorable state.

12. Wet night and morning fine afterwards – put a new wire on the machine & preparing for making a large paper for Hunt & Co.

15. Began making the large paper yesterday but did not do much. The cutter gave out. Today also we were troubled with the cutter till dinner time when we found out the cause – since which it has done very well. James went to Alton today to try to get some Rolls to glaze our paper more.

18. William went to Alton today to ask if Mr Spicer had a breast roller to spare.

DECEMBER

6. [Tuesday] The large paper we made is sent in and approved except that it has too many small specks in which for making some more for which we have an order we must if possible avoid and I trust we shall. I thought this morning how we had been labouring to do it ever since we had the machine for two years.

17. James purposes going to London on Monday and from thence to Shepton Mallet &c. He wishes to see Mr Coles Mill and Mr Yeeles, both to let.

> Joseph Coles was the paper-maker at Lower Wookey Mill, downstream from Wookey Hole near Wells. St Catherine's Mill near Bath, was also known as Mr Yeeles Mill and George Yeeles was the paper-maker at the nearby Trevarno Mills in the 1830s.[5, 26]

24. James returned safely. Went to Bath and from thence to Wells and look'd over the Mill a very complete place and in very good repair, every thing done so well at a great expence. The Rent including a corn mill and 30 acres [12.1 hectares] of 500£ pr ann – which will reduce the rent under 400£. He did not go to Mr Yeeles.

1843

JANUARY

30. [Monday] Mr Maybanke talked with James about the Mill at Whookey Hole he said a friend of his had some money he wished to employ and he would ask him if he would take part with James in it.

FEBRUARY

3. [Friday] Rcd a note from Mr Walthur Bookseller in Picadilly saying he wanted 100 or 200 Rm Demy. We sent sample sheets by post.

9. Trade continues much the same perhaps a little more sale.

20. Charles Harding turn'd off today. When I was down on Friday the cover of the Engine was off from the stuff getting under the roll from being left – this morning he would not get up and when he did he drove the Engine so fast as to endanger the tackle being broken to pieces and left the Engine to run over and waste the stuff.

> Harding had been bound apprentice to Simmons in March 1839. He had been in trouble previously in December 1839 and in September 1841.

MARCH

23. [Thursday] James went off to Worthing and call upon our rag collectors. Abm Harding today in stepping on a stool with a box of Rags on his back fell down and broke one of his ribs.

MAY

17. [Wednesday] Willm went to Godalming for Jas who return'd safe – found trade still very dull – could get no orders.

31. Mr Chorley came over with a man to set the tackle at the New Mill to right.

JUNE

9. [Friday] The weather continues exceedingly wet but having some Millwrights this week at the New Mill we are not able to do much work – trade still continues very bad.

JULY

26. [Wednesday] Jas returned from London – trade was dull.

AUGUST

8. [Tuesday] I am still as to my wordly affairs much in the same state – disappointments trouble me much – I cannot sleep.

21. Mr Spicer from Glory Mill came today.

This was James Freeman Gage Spicer of Glory and Hedge Mills at Wooburn, Buckinghamshire, whose father Freeman Gage was first cousin to John Edward Spicer, snr., of Alton.[22]

23. Mr Spicer and James went to Alton.

SEPTEMBER

27. [Wednesday] Trade very dull no orders.

OCTOBER

12. [Thursday] We have it in contemplation to have our Glazing Rollers turned and polished and to change the places of the large cylinder with the smaller one and have that also turned & polished and run with the other rollers to get a better surface. The expence is considerable as time lost.

20. Went to London. The trade very bad although trade is beginning to mend in the manufacturing districts.

23. Mr Howard came in the evening from Buckinghamshire.

28. Our friends left after breakfast. Mr Howard had not seen the place since his Father left it about 33 years ago.

Watermarks used by John Howard, snr., when he was tenant paper-maker at the Haslemere Mills (1801-11) include the only known use of the County name 'Surry'.[29]

NOVEMBER

1. [Wednesday] Sent one of the Machine Cylinders to Midhurst to be turned and polished from glazing.

11. Mr Mintley has been here today putting up the suction apparatus.

15. Sent the two Glazing Rollers to Mr Chorley, we are now quite still – at the present time I make these little improvements which are quite necessary with hesitation as every little expenditure is of consequence to me.

25. James went off yesterday for Mr Spicer's Glory Mill.

DECEMBER

6. [Wednesday] Returned from London. No trade.

15. Mr Chorley and two men came yesterday and began about removing the Machine Rollers – today we have raised the large one to its height. I am very anxious to get on again although we have no orders of any consequence to get on with – trade is still lamentably dull and things go very contrary, but it is the Lord's will that it should be so and I am desirous to bow in submission to it.

27. Sent to Midhurst for the Glazing Rolls.

1844

JANUARY

3. [Wednesday] James returned this evening. Trade very bad – prices lower and no orders. I have sold no paper of any amount for the last six weeks – other cash I ought to have received some time ago do not come and I have bills to honor next week. James sold but little and that at very low prices.

22. Making large Plate Demy.

25. Returned from London without selling a single ream of paper (only to send 2 Rms as sample) which to the best of my recollection I never did before. Yet I never had more need for selling.

MARCH

7. [Thursday] Mr Fox brought me to London. Met Jno Newman from Dartford and went with him to Mr Mackintosh, who showed us all his printing establishment. I had never seen the Machine printing before and was gratified with it. I did today get rid of a few papers at a low price and made out in regards to my Duty & acceptances better than I could hope for.

> Edward Fox, a silk mercer of Snaresbrook, Wanstead, Essex, was married to Emma, sister of Thomas Lunnon, jnr. Simmons's nephew John Newman was apprenticed to Bryan Donkin at Dartford in 1839.[5] The printing works of Alexander Mackintosh were at 19 and 20 Great New Street.

12. Mr Wyatt brought word that the Compton windmill was burnt down yesterday morning. George Newman rents it, he was there at the time, the winds very high, he tried to stop it, but could not and the friction was so great that it ignited and burnt it down. He was insured and so was the Landlord.

> Several George Newmans are mentioned in the diaries. This one appears to be the son of Simmons's sister Catherine. He had been apprenticed to Henry Roker, miller of Hatch Mill at Godalming, in 1836 and became tenant of Compton windmill, six miles south of Petersfield, in 1841.

APRIL

11. [Friday] Last night about 10 oClock just as I was going up stairs I heard a noise and cry of fire. I put on my shoes and ran out and saw the corn Mill on fire. We all ran down and when I got there the fire was through the roof of the Mill and no chance of saving it. Nothing of the Mill was saved but the waterwheel. We came away about 3 oClock – Today the men went & pulled down the front wall of the Mill which appeared dangerous. The Mill insured for eight hundred pounds. Furniture & stock 200£.

> This was Shotter Mill where George Oliver was now the miller.

MAY

18. [Saturday] James was in London this week but could sell no paper.

JULY

5. [Friday] Took stock this week; this year has not made our circumstances better, yet not so bad as we calculated on. I am still at a loss how to act but I have some kind friends to advise with.

AUGUST

16. [Friday] The Team gone to Storrington today for Rags.

21. Mrs Simmons called on Mrs Barlow. She was my first acquaintance at Midhurst when I went there to school now more than 50 years ago.

> Little is known about Simmons's early life so this entry is valuable. It seems likely that he met his wife's brothers at Midhurst Grammar School as the Newman family lived at nearby Cocking. Later Simmons and his two sisters Catherine and Ann married Charlotte and her two brothers Anthony and Thomas.

SEPTEMBER

10. [Tuesday] A prospectus is out for a Railway from Epsom to Portsmouth to come near Haslemere.

> This was the Direct London & Portsmouth Company's scheme to extend a proposed Croydon to Epsom line through Dorking, Godalming, Haslemere and Petersfield to Portsmouth. It was to be an 'atmospheric line' in which a piston contained in a continuous pipe between the rails of the track was propelled by pumping air from the pipe. The piston was connected to a coach of the train by a rod passing through a slot along the top of the pipe.[41, 42]

OCTOBER

18. [Thursday] Went off early for London. Trade very dull with us but of late have been working on low order.

24. Some surveyor of the intended Portsmouth Rail Road direct came to us today, the present survey takes all the buildings north of the dwelling house.

DECEMBER

13. [Friday] James brought from Haslemere several papers of the Specification of the Land &c. that the proposed Railway would pass through in my estate. I am decidedly for it as a public good yet my Estate will be much depreciated in value by the manner in which it passes through it.

14. James purposes going to London to attend a meeting of the paper makers & stationers to consider the propriety of petitioning Parliament to take off the Duty on paper.

> The Import Duty on rags and pulp was abolished in August 1845 but Excise Duties continued until October 1861.[12]

1845

JANUARY

1. [Wednesday] In looking back to the last year at this time I see I had a stock of paper & could sell none nor got any orders. This year though prices are very low I have had orders and sold a great deal of my stock yet I fear not to the improvement of my circumstances.

MARCH

12. [Wednesday] James met me and after breakfast we started off for Mr Spicer, Wooborn Bucks – looked at the Town Mills.

13. Looked at Mr Spicers Stock then drove over to Wickham. Called on Mr Hearne and look'd over his house & shop. Called on Mr Venables who was from home.

APRIL

4. [Friday] James had an offer of going to Spain or Russia to superintend some Mills there – rather advantageous in a pecuniary point of view, but on consulting his friends he has declined them.

> By 1851 Bryan Donkin had made and erected 14 paper-making machines in southern Europe and one in Russia.[34]

MAY

1. [Thursday] Return'd from London. More sale for paper but no improvement for price.

12. Mr & Mrs Trimmer called with Captain (Now) Sir Willm and Lady Harris to see the Mill & machine, but it was not at work.

24. James has been into Buckinghamshire & found that the Town Mills are let to Mr Venables.

JULY

26. [Saturday] The Direct London and Portsmouth Railway has passed the Committee of the House of Lords, so that we now expect it will come by us. James was in attendance all the week but not called.

> A successful trial of the atmospheric line between Forest Hill and Croydon was held on 22 August 1845, the train exceeding 60 m.p.h.[42]

AUGUST

4. [Monday] Mr Mellersh called today and left a su[bpoena] for me to attend at Croydon as a witness in a trial, George Oliver and Jno Lucas plaintiffs and Willm Shaw defendant – respecting the Insurance of the Corn Mill which was burnt down.

12. Saw Mr Mellersh in the morning when he told me the trial was referred to Council.

25. James went off this morning to attend a sale at Frogmoor Mills near Rickmansworth there being two or three cast iron Engines for sale.

Donkin's first paper-making machine was erected at Frogmore Mill in 1804.[2, 34]

SEPTEMBER

27. [Saturday] My time in the paper making concern is very nearly closed & James takes to it before long. I have been in it from (I think) Michaelmas 1811.

OCTOBER

16. [Thursday] Dear James married today.

James IV was married to Ann, daughter of the late Thomas Lunnon, snr., and sister of Thomas jnr., at Wooburn, Buckinghamshire. Simmons and his wife did not attend but his children William and Catherine did. Unfortunately William's clothes did not arrive until after the ceremony. The honeymoon was spent at Llangollen.

NOVEMBER

21. [Friday] Men at work in the Mill taking down the stone supports by the side of the machine & putting up posts to support the building to make more room to get at the Machine on the millside.

DECEMBER

2. [Tuesday] Went to London as a witness in the trial of "Oliver versus Shaw".

27. Dug out the drain on Critchmor that brings the spring water to Pitfold and new laid it in.

Figure 12. Sickle Mill from the west in August 1850.

The mill house displays the new verandah referred to in the diary entry for 5 June 1847. A prominent feature of the mill buildings on the right is the vertical shuttering of the drying lofts. Simmons recorded the erection of the smoking chimney in front of these buildings on 3 November 1840. The blurred shaft behind the house was added later in pencil and was probably associated with the new steam engine started according to the diaries on 18 February 1854. The tail water of the mill is seen at the left and the fence in the foreground follows the boundary between Haslemere and Linchmere. From a pen and wash sketch by John Wornham Penfold, jnr. (Courtesy W.J.D. Cooper)

1846

JANUARY

3. [Saturday] James Busy yesterday and today taking out the Engines on the second side to make room for some new cast Iron ones.

20. The team went today for one of the Cast Iron Engines.

FEBRUARY

5. [Thursday] Made a dry wall by the side of the pond below the new drying house and brought the waterway under the Sizing House further into the pond.

MARCH

4. [Wednesday] Mr Fourdrinier was down with James yesterday about a machine called a Willow from tareing the Rags to pieces, nearly with hemp.

> Joseph Fourdrinier was the eldest son of Henry Fourdrinier after whom the paper-making machine was named. He was trained as a paper-maker at Two Waters Mill, Hemel Hempstead where the first successful machine was installed. Later he managed an engineering works in Hanley, Stoke on Trent, making parts for machines.[43] Sale particulars of 1854 state that there was a 'Willow Room' at both Sickle and New Mills.[8]

7. Mr G Magnay came from Liphook with James and looked round & had some lunch and went on by Coach.

1847

JUNE

5. [Saturday] James is putting up a Verandah at Sickle Mill he has new stucc'o'd the House and is pulling down the Rag Houses &c. and the fuel House behind.

> The new verandah is shown in the 1850 drawing of Sickle Mill from the west by John Wornham Penfold, jnr., which is reproduced as figure 12.[27]

14. Cleared out the old counting House today and left it previous to its being pulled down, with the Rag House Woodhouse, Boathouse and also the Turf House opposite. The tiles sold to Teasdale for 12£ the Timber to Aylwyn for 25£ James keeping the boards and lead.

1847

JULY

13. [Tuesday] Exceedingly busy, in making all things strait and decent for the marriage, putting up the fences where the old rag house was pulled down. Mr Cooper and his two sisters came to tea.

14. Dear Catherine was rather nervous during the ceremony. We sat down to about 34 or 5 and everything went off very comfortably. Mr Cooper & Catherine starting off about three oClock.

> James III's daughter Catherine aged 34 was marrying William Cooper, a wholesale stationer and wallpaper manufacturer of London. The wedding was intended to be held at the new Shottermill church but Simmons discovered at the last minute that it had not been licensed for marriages and the service had to be transferred to Frensham church. The honeymoon was spent in Paris and the Coopers went to live at Holloway and later at Bickley.

18. This most probably will be the last Sunday I shall spend at Sickle Mill as we purpose moving this week to the New House to be called Field End. I came here with considerable property. I leave with very little yet I cannot regret it for myself for the Lord has been most merciful to me a sinner.

> Field End (NGR SU889328) was located at Lion Green, which is marked on the map of figure 1 about 300 yards (274m) north of Sickle Mill. The house has been replaced by council offices and a car park. It is illustrated in figure 13.

24. The week very busy in removing to Field End. James I expect will come to Sickle Mill on Monday for good. He is making a passage through the House. I fear it will be rather expensive.

AUGUST

17. [Tuesday] Mr Moline, & Mr Donkin Mr Kidd came to fishing and drank tea with us.

> Moline and Kidd were bankers of Godalming although the Kidd family were also millers. Bryan Donkin was now aged 79 so this was probably his son.

31. A meeting of the Directors &c. of the Direct London and Portsmouth Railway was held last week; it was settled to make a beginning.

SEPTEMBER

2. [Thursday] Heard a different report of the Railroad — that there could not be money sufficiently raised and that it must be abandoned altogether.

> In practice atmospheric railways had been found to have many technical difficulties and sponsors were disillusioned. In any case the period of 'railway mania' when funds poured in for any railway proposal was over.[41, 42]

4. I shall want to be Haying and also a day at Mr Sweetapples to value the stock at the Mill.

> Thomas Sweetapple was still at Catteshall Mill, Godalming (see 2 September 1831) and remained there until 1865.[16]

16. Went to London and done some business for James.

OCTOBER

19. [Tuesday] James went to London; trade very dull.

1848

MARCH

21. [Tuesday] James is gone to London, trade is very bad and distresses him much and me also.

29. The confusion and disturbances in most of the States of Europe has a lamentable effect on trade. Dear James is much distressed by it and consequently so are we.

> During early 1848 there were revolutions in many European countries including the proclamation of the Second Republic in France on 25 February, rioting in Vienna leading to Metternich's resignation on 13 March, the adoption of new laws in Hungary on 15 March providing for responsible government from Budapest, riots in Berlin from 15 to 21 March against the merger of Prussia into Germany, five days of street fighting in Milan on 18-22 March forcing Austrian forces to withdraw and a proposed Chartist demonstration at Kennington on 10 April in response to which the Duke of Wellington was entrusted with the defence of the Capital.[37]

APRIL

30. [Saturday] Trade very bad James is perplexed and cannot tell what to do for the best and I cannot help him.

MAY

17. [Wednesday] Robt Salmon called. Mr Pewtress talks of shutting down the Mill but he has not as yet, trade is very bad with him as well as with James, he is very much worried with it. Many years ago my returns were above 1000£ per month now at the Mill not more I think than 200£ or 300£ and in my own concern 20 or 30£ is of great consideration with me.

Figure 13. The south facing garden front of Field End, the house where James Simmons III lived from 1847 until he died in 1868. The diary entries for 3, 5 and 7 July 1866 state: 'took out the window and put in the girder.' 'Laid in the foundation for the Bow window.' 'We did not like the appearance of our bow window.' (Wood-engraving by Michael Renton)

78

18. Mrs James had a letter this morning from Miss Howard to tell her of the death of Mrs Acton – she had I have no doubt a great deal of property to leave and I trust the Lord has directed her so to dispose of it that there will be no room for contention and division about it.

'Mrs James' was James IV's wife Ann (*née* Lunnon) and Miss Howard was a companion of Mrs Acton and perhaps related to John Howard paper-maker of Wooburn. Mrs Acton was a wealthy and influential lady with houses in London and Buckinghamshire, and was a relative of the Lunnons. Potentially the Simmons family, particularly Ann and hence James IV, would benefit considerably from her estate.

25. James returned from London last night, he saw Mr T Lunnon after the funeral yesterday. All the relatives have a share except Mrs Fox (this I grieve over) also legacies to the servants – and also (unexpectedly) 200£ to myself – 100 to William 100£ each to my sisters Mrs A and Mrs T Newman (widows). I have to be most thankful to the Lord for his mercies to me – this little will be a great help to me. And James also will be brought out of many difficulties from what is left them.

By this time Thomas Lunnon, previously attached to James III, was probably the master paper-maker at Hedsor Mill, Wooburn, Buckinghamshire. Mrs Fox was Emma (*née* Lunnon) the elder sister of James IV's wife Ann. Mrs Anthony Newman and Mrs Thomas Newman were James III's sisters Catherine and Ann.

30. James has not as yet had any official intelligence of what is left to him or his wife by Mrs Acton – he understands that it is in Houses and that the rent do not commence for more than twelve months after her death. I am thankful for what she has left to me, but if I had sold the Estate & Mills at the time it was in contemplation, they would have sold for much more than the 200£ than now, but as she expressed a wish they might not be sold it was acceded to.

'Contention and division' (see 18 May) about the way in which Mrs Acton had left her property was already surfacing!

JUNE

7. [Wednesday] Mr Lucas from Buckinghamshire is at Sickle Mill looking over the premises with a view of taking them in connection with Mr Lacy. I think it will not come to any thing.

OCTOBER

10. [Tuesday] We have to be thankful for bodily health but are tried with other troubles so as to be at a loss what to do for the best trade being so bad James will lay still the Mills, & then no rent coming for him to pay his way and no cartage to employ my horses.

20. James returned from London – trade still very ruinous. A Scotch Maker sent 12000 Rms into the Market and sold it for 5¼ [d] pr lb a good paper. Messrs Hodgkinson & Burnside bought it, and sold again at 5½ but they have now raised their price.

> Simmons clearly considered the price of 5¼d. per lb. which included the Excise Duty of 1½d. to be too low. Hodgkinson & Burnside were wholesale stationers of London. William Sampson Hodgkinson acquired the Upper Mill at Wookey Hole mill near Wells in 1858 and his family continued to produce high quality hand-made paper there until 1951. The mill is now a museum but still demonstrates making paper by hand.[44]

NOVEMBER

10. [Friday] The working of the Mills is drawing to a close. James expects a Gentleman to look at them in a day or two either to rent or to purchase. I am very anxious that either one or the other should be done with them that some rent may be forthcoming to pay those who have a claim on them.

18. Mr Somerville, a paper-maker from Scotland has been to look at the Mills, he thinks it the best water by far that he has seen – but the power is not so much as he requires, but he will consider it over and compare it with other offers he has made to him, and then decide.

> William Sommerville had been in partnership with his father at Dalmore Mill, Penicuik, near Edinburgh, and went on to establish the Golden Valley Mill at Bitton, near Bristol. Paper was made by Wm. Sommerville & Son at Penicuik until the 1990s.[45] The 1854 sale particulars of Sickle Mill state that the washing water had been analysed by Dr Wilson of Edinburgh and pronounced the purest water he ever tested.[8]

1849

5. [Saturday] Mr Lightfoot Junr came and James went with him to Chichester and Portsmouth, he has gone back again to Newcastle to see about coal. Mr L Senr is also gone. I think he is very anxious thinking it may not answer his sons purpose, and would prefer not putting up the Steam Engine at present; but that is the very thing that in my opinion will make it answer.

> Presumably there are entries about the Lightfoots in the missing diary covering the period November 1848 to April 1849. John Lill Lightfoot was officially reported as the new occupier of Sickle Mill on 20 May 1850.[5] In 1775 a Richard Lightfoot of Dublin was supplying laying and sewing wire to paper-makers.[3] Clearly, since the installation of the paper-making machine, the water power available had been inadequate.

12. Both the Mr Lightfoots left early in the week and have not heard since.

15. James had a letter from Mr Lightfoot saying he could not arrange his pecuniary matters and would he feared be oblged to give up the concern, which very much surprised us – as he had bought Rags and some of them came in, had arranged to go to the cottage for a time till his wife could come and James continue at Sickle Mill. I think he has been told of other Mills at a low rent and wishes to make a better bargain here. It puts us to much anxiety as we had expected it was all settled and that they would have begun immediately.

19. We hear nothing from Mr Lightfoot.

24. James had a letter from Mr Lightfoot to say his partner refused to come forward in this concern and he had not sufficient without it to carry it on – therefore he had no alternative but to give it up. This make it most distressing to us, it sets aside all our arrangements and puts us in a very unpleasant position both as regards ourselves and others. It appears on the part of Mr Lightfoot that he engaged to take these Mills without having made the necessary enquiries as to the Capital required as he says in his letter that the credit is much more than he thought of and would require 1000£ more than he calculated on.

JUNE

8. [Friday] James met Mr Lightfoot Senr in London, his son wishes to take the Mills himself without Mr Findley as he and his friends are now averse to it.

JULY

12. [Thursday] A letter this morning to say that Mr Lightfoot Senr would come over early next week to start the Mills.

21. Mr Lightfoot Senr has been down and intends to remain, but went to Portsmouth to settle for Rags bought there.

28. Preparing at Sickle Mill for starting the Mills. Mr Lightfoot Senr here, his son coming next week.

AUGUST

2. [Thursday] Mr Lightfoot began working the Machine today.

DECEMBER

6. [Thursday] Mr Lightfoot cannot get any money as yet from his late agents effects — and consequently it makes him very short and which puts both James & myself to great inconvenience.

1850

JANUARY

19. [Saturday] I fear it is not well at Sickle Mill for Mr L c tells me he shall be a great loser by his late agent in London besides some loss at the Mill.

MARCH

9. [Saturday] James had some conversation with Mr Lightfoot about the Mills & the lease, he said he hoped he should go on very well after a little time and would pay the rent monthly and the carriage also. I fear if we shall find it so, as they have not money sufficient to carry it on to advantage.

Lightfoot is using Simmons's wagons to transport his paper.

15. Mr Lightfoot and his wife leaving Sickle Mill for the North for a time. I do not like appearances, and have written to James to say so. Mr L – Senr is to remain here and manage the concern which I fear will turn out a bad one for want of capital to carry it on.

27. At Sickle Mill I fear things go on sadly – an execution was put in last night by Mr Long of Portsmouth.

> Robert Long was the rag & rope dealer of Portsmouth, with whom Simmons interacted in 1837.

30. The Excise has also put in at Sickle Mill for duty. Affairs with Mr Lightfoot are very bad. The more so from want of principle, no bills are paid; no money in hand and not more than enough to pay the duty.

APRIL

9. [Tuesday] The sale at Sickle Mill. James attended.

13. The Lightfoots are all gone and Sickle Mill left unoccupied, on making up the accounts there is but little left for Rent after paying duty and expenses (about 20£).

20. James has been with us pretty much this week arranging matters about Sickle Mill. Some paper that was in the Waggon when the Execution was put in I had drawn up & put into my Granary, having a lien on it for carriage – when the sale was Mr Long would have it sold, as being off the premises, he had a right to it. Mr Pimm to whom it was consigned in London claimed it as his. Mr Lomas consequently sold it on the sale day and we bought it, & it now remains to be decided who shall have the proceeds, we have petitioned for it to go to the Excise duty, but we have not heard the decision of the board – but expect as they have got their money they will not interfere. James went off today, but will return next week, he wants the License transferred so that he may work up the stuff left in the Chest & Engines, which he bought at the sale.

> This extract was featured in a local newspaper article in 1985 and led to the discovery of the diaries.[46]

22. A Note from Mr Lomas to say he must pay the am't of the Paper sold, which was in my possession to Mr Long, and wished to know my claim on it. I said 10s/.

26. Mr Pimm continues to put in his claim for the paper; but the more I think of it the less his claim seems to be. He has only acted as Agent to Mr Lightfoot and I have no doubt but this was sent to him in the same way & that he would have his commission on it and credit his employer for whatever the paper sold for. Mr Lightfoot sent his balance sheet today – all the money he has entered when he came here for carrying on this concern was 270£ – his debts contracted since he has been here about 870 – no assets except losses and expences.

27. A letter from James he has to attend the meeting of creditors of Jno Lightfoot.

MAY

4. [Saturday] James has been waiting for the transfer of the Licence fr[om] Mr Lightfoot that he may work up the little Stuff left on the premises and which he bought at the Sale, but like every other transaction of theirs, they promise but do not perform.

JUNE

5. [Wednesday] I wrote to the Archdeacon to say that Mr Candy had arranged for a prayer meeting at Field End, not only for the District but individually also – For we were in trouble for our tenant had left the Mills without paying any rent &c. and they were now still.

AUGUST

10. [Saturday] The Mills continue to work for Mr Pewtress, which if he can keep them on will make some Rent to pay my sisters with. Mr Penfold is very kind about it, and does not complain although it must be trying to him as well as to my sisters, and a great burden to us also. We think of advertizing and trying to sell them.

> This is the only direct reference in the diaries to the Pewtress family working the Haslemere mills and their presence must have been short lived. James III's sisters, including Mary Penfold, received annuities based on the profits of his business. The views of Sickle Mill reproduced as figures 8 and 12 and that of New Mill shown as figure 14 were all drawn by Mary's son, John Wornham, jnr., in August 1850.[27]

SEPTEMBER

16. [Monday] We expect Mr Fourdrinier with another Gentleman tomorrow evening to come down and look at the Mills we were so deceived by our last tenants that we must be very careful who we now have.

> This is Joseph Fourdrinier, whom James IV met on 4 March 1846.

18. Mr Fourdrinier with Mr Smith came last night and was down early to Sickle Mill this morning – Breakfasted with James, liked the country, agreed as to the Rent of Mills but would not take to or pay for any machinery, but would [pay] some rent for it. All appeared very fair and open and after some enquiries we are to send up the particulars and they will decide immediately. Mr Smith is in the law, a man of property who will advance sufficient capital so as to pay money for every thing.

21. James is gone to London today to enquire about Mr Smith &c. and see them also about the Mills.

30. We hope and think that Mr Fourdrinier will take the Mills; although there is a point or two undecided.

OCTOBER

5. [Saturday] Some letters have passed between Mr Fourdrinier and James. Mr Smith who helps the former get into the Mill will not agree to pay for any Machinery or Fixtures, but only for the few utensils – but will pay interest for it at the rate of 6 prCent. If we did not want the money the interest would be a fair compensation.

8. James having written to Mr Fourdrinier saying we would accede to his terms, he came down this afternoon and I trust it will be settled very speedily, it was 8 years ago this day that the machine was first set to work.

> According to the diary entry of 22 March 1841 Simmons had already started making paper with his machine.

NOVEMBER

9. [Saturday] We have heard nothing of Mr Smith or Mr Fourdrinier, and are anxious about it.

16. We hear nothing from Mr Fourdrinier.

DECEMBER

11. [Wednesday] James rec'd a letter yesterday from Mr Fourdrinier with the draft of the lease, with Mr Smiths observations upon it; he wrote wishing Mr F – to come down.

14. Mr Fourdrinier came down on Thursday, he seems to agree to the terms of the lease, he went up last evening to see Mr Smith this morning, and hopes to come down the latter end of next week. We heard yesterday that Abm Harding (son to Mrs Smith) who went over to America and married just before he started lost his wife, about a month after he arrived there.

> See the note for 27 February 1835 for further information about this member of the Harding family.[19]

Figure 14. New Mill from the north-east in August 1850.
The extensive shuttered drying lofts are the main feature of the building and the new water wheel installed in 1840 is clearly seen at the left. The mill pond, which appears in the foreground, has now been filled and the mill, with its shuttered drying lofts, was demolished in 1976. From a drawing by John Wornham Penfold, Jnr. (Courtesy Haslemere Educational Museum)

1851

JANUARY

4. [Saturday] Mr Fourdrinier & family came today to Sickle Mill.

8. Mr Fourdrinier wants many things done at the Mills and talks himself of putting in a new cutter and also a Steam Engine to drive the machine.

11. Mr Fourdrinier told me yesterday that he wish'd to have the Bill for carriage the last Saturday in every month & the following week I should have a check for the amount.

18. Mr Fourdrinier has begun making small hand very thin and the machine doing it very very well.

21. Mr Fourdrinier is having the spur wheel at the New Mill new geered, when that is done all the wheels at both mills except the water wheels will be in good trim.

23. Went today with the Horses to bring down a new cutter to the machine, it is expected that Mr Warren will buy the one taken out.

> The fact that William Warren of Bramshott, Barford and Standford paper mills needed a cutter indicates that he had also installed a paper-making machine. See 17 December 1840.

28. The paper trade are making a great stir to get the duty on paper taken off; but I think it is chiefly with the publishers, & not so much with the manufacturers. I think myself there are other taxes of more importance to the country that might be done away with. If they had only asked for the 5 pCent which was put on a few years back – that would have been in the makers fav'r as it was quite out of their pockets when put on.

FEBRUARY

1. [Saturday] I have this week taken the first money for carriage from Sickle Mill. I am very thankful for it and the arrangement is that it is to be paid every month. Mr Fourdrinier has been very busy with the workmen in cleaning out the stream below New Mill, new geering the fly wheel and raising the Tackle 9 Inches [230mm], putting the Water Wheel up stream and adapting the fall to it.

MARCH

1. [Saturday] Mr Fourdrinier raised the Water Wheel &c. at the New Mill and New geered the Spur Wheel and it goes very well and hopes to drive two Engines comfortably – he has also put in a Boiler to heat the retort with hot water – it seems to answer very well.

> Later sale particulars (see 19 October 1854) give further details of the water-wheel and indicate that the retort was for bleaching.

13. Mr Smith came to Sickle Mill yesterday & returned this morning. Mr Fourdrinier set on the machine on tissue paper today – it does very well and he seems pleased with it.

> An account of paper-making at Sickle Mill, as recalled by Joseph Fourdrinier's daughter who was aged about 9 at the time, has been published. In particular she remembered seeing beautiful white tissue paper coming down the cylinders of the machine and packets of paper being sent off in waggons by road to Henry Fourdrinier the wholesale stationer. Joseph's brother George also made tissue paper at Ivy House Mill, Hanley, Stoke-on-Trent.[43]

22. The Water Wheel at Sickle Mill so broken I fear it cannot be mended.

26. Mr Smith has written to Mr F – to say he will give up the concern here so far as he has to do with it; without we will put in a New Water Wheel and all that he requires beside – We had agreed to find the Timber and I hoped all was settled but Mr Smith writes in a very preremtory and decided way. We hoped and had thought the wheel would do very well for another year when there would be time to get the Timber ready.

29. James has seen Mr Smith & I hope all things will be settled satisfactorily.

APRIL

12. [Saturday] James went with me to look out some Beech Timber for the Water Wheel set out about 5 Load which will be nearly enough – but we do not know if it will be wanted or not as we have given Mr F the offer of paying him for the wheel if he should prefer putting in an Iron one and pay the extra himself.

> Timber waterwheels were usually constructed of oak with the floats or buckets made of elm. However green or unseasoned beech could also be used and it appears that the trees belonged to Simmons so that this was a cheap solution. In practice the option of installing an iron wheel was not adopted as when Simmons sold the mill (see 19 October 1854) the particulars refer to a powerful nearly new wooden overshot waterwheel, 17 feet (5.2m) in diameter and 7 feet (2.1m) wide.

14. Mr Spicer has been very busy, his concern is large and he has been making some alterations so he (Jas) has been there assisting him.

> This is James Freeman Gage Spicer of Glory and Hedge Mills, Buckinghamshire, who was soon to be in financial difficulties.

MAY

10. [Saturday] Cut down and drawed to the saw pit this week nine Beech Trees. The South water wheel at Sickle Mill.

31. The week fine & dry, busy one day & part of another taking out the old waterwheel – getting out the Iron and sending it to Midhurst for the new one.

> Simmons is sending the iron gudgeons, which were mounted in each end of the timber shaft of the waterwheel and rested in the bearings, to Robert Chorley the millwright at Midhurst (see 7 April 1852).

AUGUST

8. [Friday] The new water wheel (South side) finished on Wednesday and has been since working and goes very well.

23. Mr Fourdrinier has been in London the last two days and has not returned – I fear all things as regards the business is not going on well – I am sure it cannot answer unless he has a better supply of Rags and the water made the most of.

OCTOBER

10. [Friday] On Wednesday I went into London with Mr Cooper and called on several of my old friends – Dined at Smithfield, I looked at their papers in the warehouse.

> Simmons is visiting his son-in-law William Cooper who owned a large wall-paper factory in West Smithfield.[2]

20. A letter from James in which he says he hopes Mr Spicers affairs will be settled more to their satisfaction than they once expected, a letter from William also. He tells me that the other night Mr Venables steam boiler burst.

> James IV and William were at Glory Mill, Wooburn. The Venables family were operating several mills nearby including Lower Glory which is probably where the steam boiler burst.[5]

27. A fine day and mild but rather gloomy to me being disappointed in Mr F – not paying the money he most faithfully promised to do – I fear it is not going on well with him.

NOVEMBER

1. [Saturday] We are straightened in not getting our Rent according to agreement.

14. James had a letter from Mr Daintrey to say the promise of the rent being paid on Saturday week was made & wished to know whether we would wait that time. I wrote to say we would.

26. Mr F – promises payment of Rent &c. but do not perform which makes me uncomfortable.

DECEMBER

20. [Saturday] Mr F says he cannot stay on at the Mills any longer for water of more power – this summer is the driest I ever remember, no rain of any consequence, not sufficient to increase the power of the Mills.

27. Hard frost. The Mills are still and the water running by – some more stuff to bleach and paper to make.

1852

JANUARY

3. [Saturday] James & William left about one oClock for Wobourn, the former wishing to see Mr Fourdrinier at Godalming. The Mills being still for the present and my Horses not employed, I talked to James about letting the land and selling off my little stock and we came to the conclusion that it would be best to do so. And as James thinks of staying at Wooburn for a time we thought if I could let Field End and go to Cherrimans it would be better but this depend on circumstances.

> Simmons had lived at Field End since July 1847. Cherrimans (NGR SU885326), which still survives as a substantial house in Liphook Road, was James IV's new home and he lived there until his death in 1903. Across the road is Brookbank, which George Eliot rented from James IV for three months in 1871 and where she wrote much of *Middlemarch*. After her lease expired she stayed at Cherrimans for a few weeks. In 1866 James IV toured the Haslemere area with Alfred Tennyson, helping the poet to find a site for his new country house.[47] This was built on Blackdown, two miles south-east of Haslemere, and called Aldworth (NGR SU925308).

10. I do not remember a drought of so long continuance, it began before haying time since which there has been no rain to make any increase of water at the Mills. Mr Fourdrinier has felt it very much, yet not so much as some Mills have.

15. William is now at Mr Spicers helping to take stock.

James Freeman Gage Spicer of Glory Mill, Wooburn, and Hedge Mill, Loudwater, was now bankrupt. John Edward Spicer, jnr. of Alton and Wandsworth Mills was bankrupt in 1846 and again in November 1852 when he was described as a paper-maker of Chilworth and Alton.[5]

31. A letter from James & William – the latter is still at Mr Spicers.

FEBRUARY

10. [Tuesday] I have been employed today in looking over papers and letters for years past, burning and tearing up a good many.

14. John Tilbury was buried today – he was apprenticed at Sickle Mill just before my Uncle James died I can remember his father coming there with him.

In 1789 James and William Simmons took as apprentices John Tilbury and Thomas Eede.[4] James II died on 10 January 1790.

MARCH

27. [Saturday] I never remember so dry a Winter nor the Summer which preceeded it. We have scarcely had a wet day since the autumn. Mr Fourdrinier complains very much of it as regards the little work he can do at the Mills, but still he has not worked what there was, not having the material to do it – had his friends who first helped him to go on and take it followed it up so as to have put up a small steam Engine as it was set out to do, he might have done very well. He now says he will do it and that the Engine is coming. I hope it may but I fear while I hope.

APRIL

7. [Wednesday] Went yesterday to Godalming to arrange for James about the money to pay Mr Chorley for the Water wheel he put in at Sickle Mill £69. 10s. – this I think includes something more than the wheel and so it should. I think Mr Chorley is rather high in his charges.

15. James started off this morning for Chichester to attend the sale of the Machinery at Ashling Paper Mill which is to be sold for a corn mill.

> West Ashling paper mill was 3 miles [4.8km] west of Chichester. It was a new mill in 1825 and had several occupiers until it was bought in 1850 by Robert Chorley, the engineer of Midhurst who did work for Simmons.[17] It did indeed become a corn mill with three pairs of stones and although it has now been converted into a house the machinery survives. Originally there was a windmill above the watermill.[48]

MAY

22. [Saturday] The Boiler today let down to its place at Sickle Mill and the foundation began laying in.

> The boiler would be for the small steam engine which Fourdrinier is about to install to power the paper-making machine.

JUNE

10. [Thursday] This and two other days this week have been very wet. Could do nothing in getting forward with the steam Engine.

14. Mr G Parson called. He said for a truth that the Railway Direct to Portsmouth was to be opened again to branch from the South Western at Godalming and join the Coast line at Havant.

> The Portsmouth Direct Railway Company was to obtain its Act in 1853 to link the London & South Western line at Godalming (1849) and the London Brighton and South Coast line at Havant (1847).[28, 42]

JULY

31. [Saturday] Mr Fourdrinier came to me on Thursday to say he had paid the Insurance, that he had arranged to pay the Rent monthly and leave it in London. I hope he will be faithful to his promise.

AUGUST

16. [Monday] Settled about Mr Fourdrinier having Pitfold Mill – Rent 60£ – for seven years – 130£ to be allowed for the steam Engine now put up out of the Rent.

31. Mr Fourdrinier told me last night that Mr Sidney Smith his former friend had put in an execution at Sickle Mill for his debt – by which he was very much distressed as it would frustrate all his hopes for the future – he went to London today and James returned with him in the evening and both went off together the next morning.

SEPTEMBER

2. [Thursday] Mr Fourdrinier not returned nor have I heard from James if any arrangement is likely to be made. I went to Petworth today to get some notices for the Sheriffs officer to retain the money due for rent.

6. Mr Fourdrinier in London. The supervisor came to Sickle Mill about the Duty. I went in to him – but could not arrange about it – wrote to James.

> Although excise officers stamped packets of paper at mills before they were despatched, the duty was usually collected several months after the paper was sold.[12]

11. The week is past, it has been an anxious one as to Sickle Mill, the Duty was paid on Friday but still nothing satisfactory arranged.

18. Mr Fourdrinier in London all the week. He wishes to try another month to show his creditors what he can do, now the steam Engine is up – a friend will let him have materials if we will not take them for Rent – which we have agreed to do.

21. Had a note from Mr Fourdrinier to say he would not be home for some time and wished his horse to be turned out also lamenting over the state of his affairs, which I fear is a very bad state – he still looks forward to going on but I see and have done for some little time no chance of it.

30. Went to Petworth yesterday to see Mr Daintrey; he advised our getting an appraiser to levy a distress for rent. I wrote to Mr Hutton the Sheriffs officer to ask him to come over, which he did and said the sale must take place the next Wednesday & if it was put off we must get Mr Fourdrinier consent, which no doubt we can obtain as it is what he wished and asked for.

OCTOBER

5. [Monday] The day of the sale at Sickle Mill is fixed for next Tuesday if nothing sets it aside.

6. Went to Petworth. Mr Daintrey says a sale of Mr Fourdriniers effects may be avoided by an assignment of them to us there not being sufficient to cover the amt due for rent at the same time notifying the relinquishment of his lease as being forfeited by his non performance of its covenants. I wrote to James from thence to that effect. I called on Mrs F this afternoon and she seems to think it would be the best arrangement.

8. Mr Fourdrinier will not make over to us the things in the Mill & house so that there must be a sale.

12. The Bills of the Sale at Sickle Mill out, the Sale to be on Friday – called on Mrs Fourdrinier she told me her husband had not mentioned about the sale that she expected that it was put off to the 24th. He wrote me an angry letter last week because it was to be so soon. If he had consented to make it over to us without a sale, we should have given them most of the furniture; or rather lent it to them because if theirs, some creditor no doubt would have seized it.

13. Mr Fourdrinier wrote me another angry letter today, charging me with breaking through my Golden Rule of Sunday correspondence to distress him and his family. I did not write to him at all it was no doubt James and I do not know it was written on the Sunday for if we write letters on a Saturday most likely we put them in on Sunday that they may be recie'd on Monday.

14. Mr Fourdrinier came home last night, it appears he does not mean to leave and to give us all the trouble he can.

15. This morning after everything was ready for the sale, he wished to see James – he had before told him that his nephew a lawyer was coming down which he did – in talking to his uncle he said why did you not comply with Mr Simmons offer and make over the things and prevent the sale it would be much better for you. He then wished to see James for to so arrange it, but it could not be done without risk of expence &c. for some brokers were come from London the people were assembled and it was too late. So the sale went on the things in the Mill sold cheap and we bought the greater part – the furniture sold very nearly what I computed it at. I bought some lots not that I wanted them but to keep up the sale, for many things would have sold at a lower price if I had not bid for them. I had some cause afterwards to regret that I did so.

16. The things all gone and Sickle mill an empty house. Mr F we think intends to keep possesion if he can, but do not appear hostile. He told James this morning that he had a friend that would pay a quarters rent in advance and if it answered would take a lease. Mr Fourdrinier intends going to London on Monday and on his return will tell us who his friend is & ask our opinion in regard to it.

18. Mr Fourdrinier went to London this morning. He does not give up possesion but agreed that Wm Voller should take the keys assuring us that he did not mean to be a rogue. I fear he is approaching very near it. I heard today that another writ is out against him.

A William Voller had been apprenticed to James III's father William in 1800.[4] This is probably his son.

23. James had a letter from Mr Fourdrinier about trying the Mills for 3 months.

NOVEMBER

9. [Tuesday] Letter from James to say he should be here tonight & set the Mill on to work up the stuff in the Mill.

11. James came the other evening preparing to set on tomorrow. Some Rags came down. Mr Fourdrinier has signed a paper to give up the lease of the Mill, but to have the priviledge for 3 months of recommending a tenant, which we have agreed to accept if we approve of him.

12. Began at Sickle Mill working up the stuff in the Chest.

16. James left us today, and will not return for a week – sorting and boiling rags, and getting forward for next week. I heard this afternoon that Eashing paper mill were burnt down.

Thomas Pewtress, James Lowe and Benjamin Pewtress of Iping Mill had acquired Eashing Mill in 1833 and the Pewtress family continued there until about 1869. A newspaper reported that the fire caused damage estimated at £5000.[5, 49]

17. Mr Pewtress called in his road from Iping to Eashing. He did not know the extent of the damage. I afterwards heard that the Machinery was not much burnt.

26. A very boisterous and wet day; we are little inconvenienced by it at Sickle Mill. Wm Wailer [Voller] has gone into the House there and Jas is getting the Rags boiled and bleached ready for to begin making next week.

DECEMBER

2. [Thursday] Began making Tissue Paper at Sickle Mill yesterday it has gone on very well for the beginning, very few breaks today.

Unlike Simmons, Joseph Fourdrinier had clearly appreciated that in order to be successful making paper in a relatively remote country town like Haslemere it was necessary to specialise. James IV had clearly accepted this principle and was continuing to make tissue paper.

8. James returned from Woburn this evening. I have been engaged at the Mills and at Haslemere since Sunday.

17. A continuance of wet weather, the Mills on the larger streams all still from floods.

1853

JANUARY

4. [Tuesday] James went off early this morning to Petersfield and from thence to Portsmouth, purposing to reach Chichester tonight, tomorrow to Brighton and thence to London. James' tour round was to buy Rags.

15. The last day or two busy at Sickle Mill, making the Old Sizing House into a Cottage. Also polishing the Glazing Rolls & making the Machine go by water as well as by steam.

18. The Machine has been still a day or two connecting again the Machine to the water wheel so that when there is a full supply of water it may be used to save coal at the boiler and not use the steam Engine.

> Having decided earlier that a steam engine to power the paper-making machine would solve all the problems associated with the unreliability of water power, Simmons has now realised that fuel for the boiler is expensive whereas water is free.

29. The drain in the old boiler house has been stopped for some time – emptied it but could not open the drain. Dug down to get it into the arch from the water wheel, could not find it & so left it till Monday.

31. Bridger came down to see about the drain, dug under the foundation of the Mill five feet [1.5m] or more could not find the arch, went up under and opened a hole and in about four feet [1.2m] reached another hole, put in some old cast iron pipes to take the water away.

FEBRUARY

5. [Saturday] James now thinks of coming to Sickle Mill again after the three months that was allowed to Mr Fourdrinier to find a tenant is expired.

9. James came to the Mill and looked at the pink paper arranged about it and left.

19. James left us this morning about noon. Drove to the Shalford Station from thence to Maidenhead.

> Shalford Station on the Reading, Guildford and Reigate Company's line, leased to the South Eastern Railway, had opened in 1849. To reach Maidenhead, James IV would have to change stations at Reading to the broad-gauge Great Western Railway.[28]

26. James has been busy taking stock, I have been helping him. As near as we can judge he made the rent and a portion over. The boiled rags and half stuff we are not certain as to quantity, yet we took it as we think within the quantity. Mr Dunsters accts agree with ours as to the materials sent down and the goods received.

MARCH

5. [Saturday] Occupied at Sickle Mill looking to the men and in making calculation & seeing to the accounts.

9. I have been busy at Sickle Mill & Pitfold. At the former Mill making a stairs from the finishing room to the Rag loft and also down to the Engine House, putting up wheels to drive a devil in the Rag House, and at Pitfold putting in a Roll & Plate.

12. Full employment at Sickle Mill &c. Put the Roll from Woking Mill into the Engine at Pitfold, get it up without much trouble and no accident. Very much annoyed at Sickle Mill this morning by finding the Boiler leaking – the cover of the Boiler burnt up and I fear all from neglect in not keeping it full. I found that it had been very hot and the leak so much that we could not stop it – on having Robt Puttick down we could get no piece of Iron of the proper kind to put on – settled to send to Guildford very early on Monday morning for it.

> Although he had retired in 1847 Simmons is clearly thrilled to be making paper again. The business must have been flourishing as Sickle, Pitfold and New Mill were all being used. At this time the master paper-maker at Woking Mill was Henry Virtue Bailey producing newsprint.[5]

16. At work with the Machine last night and today it goes on pretty well.

18. James gone to London today to see Mr Dunster and talk over the concern at Sickle Mill.

26. James & his wife sleep at Sickle Mill tonight.

28. This morning we again found some evil disposed people had broken the pipes in Pitfold Meadow so as to stop the washing water from Pitfold.

31. James very busy at the Mill taking out the old Boiler for the Rag boiler and fitting the tubes up to receive the waste steam from the Engine.

> The waste steam from the engine could be fed into pipes and hence used to dry paper. Simmons is now referring to two different types of boilers and two different types of engines. Rag boilers were used to clean and break down the rags and steam boilers were needed for steam engines to power the paper-making machine and the beating engines which produced stuff or pulp.

APRIL

2. [Saturday] James went to Mr Higginbothams about wood for the steam boiler.

9. James is gone to London to seek for a Boiler.

MAY

25. [Wednesday] We are expecting the Railway Bill to pass into the House of Lords every day.

JULY

9. [Saturday] The new boiler first used.

This appears to be a new rag boiler to replace the one damaged on 12 March.

16. The weather very wet since Wednesday. The Hay spoiling – what Jas will lose on his Hay, he has the advantage of water for the Mill and save much in the less consumption of Coal.

AUGUST

6. [Saturday] The first sod of the direct London & Portsmouth Railway was cut today at Buriton. I went with James.

Buriton is 2 miles [3.2km] south of Petersfield. The diary gives a full account of the ceremony and festivities.

SEPTEMBER

13. [Tuesday] At Sickle Mill today posting up the accounts.

OCTOBER

1. [Saturday] James & his wife gone to London unexpectedly, he having a letter from Mr Dunster telling him of a sale on Monday, where some Engines &c. were to be sold.

5. Into London called Mr Dunster and met James went with him about a Steam Engine.

NOVEMBER

5. [Saturday] Busy at Sickle Mill at the Accts James preparing for the Steam Engine he bought & in the arrangement for it. Johnathen Openshaw turned off for drinking.

This must be the '25-horse high pressure and condensing steam engine', used for powering beating engines, referred to in the 1854 sale particulars of Sickle Mill[8] (see 19 October 1854).

28. James went to Hedsor on Saturday – to see about a fly balance wheel for the Steam Engine, as he knew there was one out of use at Mr Spicers, he bought it very well. The one belonging to the Steam Engine was broken by the South & Company's men in unloading it so that they have to make it good.

> In 1852 James Freeman Gage Spicer, of Glory Mill, Wooburn, and Hedge Mill, Loudwater, Buckinghamshire, conveyed his estate to creditors.[5]

30. The last week the Steam Engine was brought at different times and the unloading it has been difficult, but thanks be to the Almighty as yet there has been no accident – today also moving the upper Engine to get it in its place for the Steam Engine. Mr Thos Lunnon had a fire, but not a very serious one; he was insured.

> Thomas Lunnon, jnr., James IV's brother-in-law, was at Hedsor Mill, Wooburn, Buckinghamshire.[5]

DECEMBER

7. [Wednesday] Very busy at Sickle Mill getting ready for the Steam Engine. James went to Portsmouth to a Dockyard Sale and bought canvass &c.

9. Yesterday & today occupied at S Mill in the same way; but not with the same success for after getting the iron fly wheel along the pond head in letting it down flat to slide into its place, by some means the rope slipped when nearly down, and it fell on Chr Bridgers foot being under it bruised very severely. Mr Clothier came immediately, he found no bone broken.

10. Lowered the Iron fly wheel into its place and began getting the cistern along the pond head.

16. Getting the Iron Cistern to its place at Sickle Mill.

17. Frost & snow. The workman came to put up the Steam Engine.

Figure 15. Photograph of Mary Penfold, James Simmons's sister probably taken by William Simmons on 8 August 1854, when he took the photograph of his father, James III, shown as the frontispiece. The original is an albumen print taken from a wet collodion plate. (Courtesy W.J.D. Cooper)

Figure 16. Photograph of John Wornham, snr., James Simmons's cousin, probably taken by William Simmons on 8 August 1854. The original is an albumen print taken from a wet collodion plate. (Courtesy W.J.D. Cooper)

1854

JANUARY

12. [Thursday] The Steam Engine all heavy parts up. Great cause for thankfulness that no other accident has happened.

FEBRUARY

18. [Saturday] This afternoon the Steam Engine at Sickle Mill made a start not to work any stuff but to get it in order for Monday or Tuesday – it appeared to do very well. It is to drive the two old engines and if we can get a supply of dry wood we think it will do it – for a short time that will be a difficulty. Coal rather lower but still even the steam Coal 30/ [per ton] at Godalming & the best 36s/. Orders come in very fast if we can make it at a profit. Rags & coal are high.

25. Jno Barker the Engineer who put the Steam Engine up, went off today. He says it is a most excellent Engine; and it works well. May we all look up to God with grateful hearts.

28. The Steam Engine going on very well but short of dry wood to keep up the steam. Coal is cheaper.

JUNE

10. [Saturday] Both Rags & Water very short.

15. We expect some Gentlemen down next week to look at the Mills, trade is brisk, but Rags so scarce and dear as not to get a supply.

20. James had business at Bramshot Mill, I went with him.

JULY

24. [Monday] Mr G Newman from Hurst came over to ask James about a Water Wheel Shaft.

AUGUST

8. [Tuesday] Mr Spicer from Glory Mill has been at Sickle Mill the last day or two.

> This is the day on which James III's son William took family 'portraits using the phrotographic art'. It seems likely that the photographs, of the diarist himself (frontispiece), and of Mary and John Wornham Penfold, snr. (figures 15 and 16), who were the diarist's sister and cousin, were taken on this day.

12. Beginning the Rail at Haslemere apparently in earnest wanting all the hands they can get.

17. Business do not go on pleasantly at the Mills. Rags so scarce that a sufficiency cannot be had to carry it fully on. Coal has been very dear, so that wood has been substituted for the Boilers, which also I think is expensive. James has offered to let them but no one has come forward. Several has talked about it but it came to nothing.

SEPTEMBER

14. [Thursday] James went & returned from London. We purpose selling the Mills by auction if we can. Sickle Mill & New Mill without any land.

30. James went to London to settle about the sale of Sickle & New Mill.

OCTOBER

10. [Tuesday] Dear William left us today. I wish he had an employment more suitable to his feelings – he feels very much that Sickle Mill is offered for sale as he hoped James would have continued it and he also assisting.

14. Expect some gentlemen down to look at the Mills on Monday or Tuesday. May the Lord grant us a sale for them at a fair price.

16. Mr Appleton and his two sons Thos & Wm have been to look at the Mills today.

Henry Appleton and his two sons Thomas Giles and William were braid, smallware and trimmings manufacturers. Henry was originally based in London but rented Pitfold Mill from James III in 1835 for about two years and then with his sons built up a flourishing business at Elstead Mill.[5, 20] His daughter Susannah married into the Pewtress paper-making family in 1840. Also, Daniel Appleton a paper-maker from Pitfold was buried at Frensham in 1790[50] and John Appleton was involved in paper-making, bleaching and logwood grinding for producing dyes near Manchester in about 1820,[26] but these may have been unrelated.

18. James gone to London. The sale of Sickle Mill & New Mill tomorrow. Mr Wm Appleton told James yesterday that 4500£ was the full value – he would go home and calculate on that amount and if his calculations agreed therewith he would make that offer before the sale. If he makes the offer I will sell.

19. The sale has taken place! The Mills did not sell at the Sale, but afterwards Mr Wm Appleton bought Sickle Mill, all the machinery included for 4450£. (The new drying house not included in the purchase). The New Mill he will rent if he does not buy it. I am very thankful that it is thus so far settled, although the parting from it is wrench on the feelings. Dear William will also feel it, as his heart was set on living in the country and having employment in a paper mill.

PARTICULARS AND CONDITIONS OF SALE

OF THE COMPACT AND DESIRABLE

WATER-POWER PAPER MILLS,

(WITH AUXILIARY STEAM POWER,)

DISTINGUISHED AS

"SICKLE MILLS,"

WITHIN ONE MILE OF THE MARKET TOWN OF

HASLEMERE, SURREY,

An easy Distance from LONDON, close to the LINE of

RAILWAY FROM LONDON TO PORTSMOUTH,

Now in course of construction, from which there is every facility for a private siding to the Mills, which comprise all the requisite

Buildings, Plant, and Machinery,

Now in full work, and supplied with the PUREST WASHING WATER, flowing into the Mills without pumps, adapted for the Manufacture of the finest Papers;

A COMMODIOUS DWELLING HOUSE,

WITH GARDENS, STABLING, ETC.

THE WHOLE COVERING ABOUT FIVE ACRES;

ALSO, WITHIN A SHORT DISTANCE OF THE ABOVE, THE

CAPITAL WATER-POWER MILL,

DISTINGUISHED AS

"HALL'S, OR "NEW MILL,"

IN THE PARISH OF LINCHMERE, SUSSEX,

Supplied with the PUREST WASHING WATER, with all necessary BUILDINGS, EXTENSIVE STORES, &c.,

Well adapted for any Manufacturing Purpose;

WHICH WILL BE SOLD BY AUCTION BY

MR. MARSH,

At the MART, opposite the Bank of England,

On THURSDAY, OCTOBER 19th, 1854, at 12 o'Clock,

Figure 17. Part of the title-page of the particulars of 1854 (x 0.57) when James Simmons offered Sickle Mill and New Mill for sale. Further details are given in the note on the extract for 19 October 1854.
(Courtesy W.J.D. Cooper)

104

The sale particulars survive and provide a wealth of information about the mills. Part of the title page is illustrated in figure 17. The details for Sickle Mill refer to the possibility of a railway siding, a good country collection of rags, trout in the mill pond, pure washing water, a powerful nearly new over-shot waterwheel 17 feet (5.2m) in diameter and 7 feet (2.1m) wide, an iron waterwheel 15 feet (4.6m) in diameter and 3 feet (0.9m) wide, gear wheels, pit and fly wheels, pinions, etc., a 25 horse power high pressure condensing steam engine, two steam boilers, presses, two large iron beating engines, and two engines lined with lead. The buildings included a machine house with small engine house adjoining, paper engine house and steam engine room, rag boiler house and store, finishing salle, counting house, foreman's store, lathe room, machine store room, willow room, rag sorting house, rag stores, fuel house, smith's shop, carpenter's shop, large drying house and foreman's cottage, and a commodious dwelling house with four large bedrooms. The site covered over 4l acres (16.6 hectares). Similarly the details for New Mill, described as a half-stuff mill well adapted for any manufacturing purposes, refer to a large pond, a powerful breast-shot waterwheel 16 feet (4.9m) high and 9 feet (2.7m) wide with 9 feet (2.7m) fall, pit and spur fly wheels and pinions, pure washing water, two rag engines, presses, and the buildings included an engine room, willow room, retort bleaching house, steeping room, drying loft and rag house. This site contained over 1½ acres (0.6 hectare).[8]

21. The day fine & mild – at Sickle Mill making out the balance of the different accnts owing &c.

23. Mr Thos & Mr Wm Appleton came to Sickle Mill this morning and settled with James how the purchase money was to be paid and it was to his satisfaction – they do not purchase the New Mill but have taken it for two years at 85£ pr year. It is great satisfaction to Mrs Simmons & myself that they are a Godly family. The father Mr A Senr is a dissenter and I believe these also; when the father rented Pitfold Mill and lived at the Cottage (Cherrimans, or rather Mr Thos A lived there) we went on very comfortably together.

DECEMBER

9. [Saturday] Wrote to Sister Sarah to day as to her signature of the release of legacy on Sickle Mill &c.

12. James gave a supper to the men and tea to the women at Sickle Mill on his leaving the concern there & Mr Appleton taking to it, he will not leave the House at present.

1857

MAY

19. [Tuesday] Went with James to Midhurst & surrendered the New Mill property to him.

20. The Railway people are very busy at the Hanger gate, putting an arch under the road.

AUGUST

3. [Monday] Gentlemen came to Cherrimans to look over the land &c. as to our Compensation from the Railroad contractors for the portion taken and the damage done to the Estate.

18. I have been troubled by the rupture & hernia & my truss I am obliged to lay by for a time. Mr Clothier thinks I had better apply leeches &c.

NOVEMBER

18. [Thursday] The Boiler putting in at Sickle Mill. Mr Appleton had it from Manchester.

> Presumably this was a Lancashire boiler, as patented by William Fairbairn and John Hetherington of Manchester in 1844.[51]

1858

MAY

26. [Wednesday] Mr Candy & Frank came in to tea. The latter showed us his invention of making netting by a machine – very ingenious but I rather doubt if it will make it in sufficient quantities in a given time to bring much profit.

29. Went with Charlotte to Sickle Mill to see the New Machinery, put up for preparing the wool for making yarn. It is beautiful machinery and wonderful to see what it can do. After the wool is prepared here it is sent to Elstead to be made into yarn.

'Charlotte' is Simmons's daughter and not his wife. Appletons were making worsted yarn and therefore the machinery would be for combing wool. In 1790 Edmund Cartwright, better known for his invention of the power loom, patented the first wool-combing machine. This was improved upon throughout the first half of the nineteenth century until James Noble patented his version in 1855 and this is basically the machine still used. The new equipment at Sickle Mill could have been early Noble combing machines.[52]

JUNE

22. [Tuesday] The first Engine came down the line today. We all went to Cherrimans to see it pass. It will now be some time before it is opened as a public conveyance. It is thought that the South western Company will buy and work it.

It had taken nearly five years to build this track, which reduced the distancce by rail from London to Portsmouth from 95 (153km) to 70 miles (113km).[28]

30. James went to London and settled partly with the Railway people for the compensation money, but not wholly, as regards the New Mill.

OCTOBER

23. [Friday] Heard this morning that Mr Pewtress son in law to Mr Appleton was dead.

Simmons recorded the wedding of Mr Pewtress and Susan Appleton on 17 October 1840.

1859

JANUARY

1. [Saturday] The day on which the Direct London & Portsmouth Rail Way was opened; went up to the Station to see the first train come in from Portsmouth at 8h. 46m a new era in our neighbourhood.

The London & South Western had indeed leased the new line but was having difficulties in obtaining an agreement to run trains over the London Brighton & South Coast track from Havant towards Portsmouth. They had sent a goods train down on 28 December but it failed to reach Portsmouth as the Brighton people had removed the points, obstructed the line with an engine and later lifted a rail. The passenger train seen by Simmons must have come from Havant and not Portsmouth as through-running did not commence until 24 January.[28]

1860

FEBRUARY

20. [Monday] James came in from London, where he had been to go with the deputation to the Chancellor respecting the paper duty, it being taken off. The French are allowed to import paper into England but we're not allowed by the French tariff to import Rags, so as to give them plenty of cheap material for making paper & while with us it would be more expensive. Mr Gladstone told them he would do what he could, so did Mr Gibson.

> At this time William Gladstone was Chancellor of the Exchequer and Thomas Milner-Gibson was President of the Board of Trade.[33] Gladstone was proposing to remove both Excise Duty on paper made at home and Customs Duty on imported paper. The deputation of paper manufacturers saw him on 18 February and wanted the latter to be retained. In practice Gladstone's Bill to remove all Paper Duties was passed by the Commons on 8 May but rejected by the Lords a fortnight later. A political crisis threatened and Gladstone's resignation was considered imminent. This did not occur however and eventually on 18 October 1861 all Paper Duty ceased.[12]

MARCH

21. [Wednesday] Today the Queen & Prince Consort returned from Osborne by rail on our line.

MAY

19. [Saturday] I was in my room after dinner for some time when I was called in; my dearest wife was awake but her breath shorter. When I said are you easy? I just could hear her say, Yes, she continued to breathe for some time but when the clock six, it ceased as quietly as if she had gone to sleep. I am now a widower; my dear wife is in happiness above forever.

> Charlotte Simmons was almost four years older than James III and nearly 81 when she died. Although her name does not appear frequently in these extracts she does of course feature prominently in the diaries themselves.

JUNE

2. [Saturday] Fifty Three Birthdays of mine are passed during which long period I have had a dear and much valued companion many indeed have been the changes of this checkered life. The prospect of worldly prosperity shone around us, as we passed through the first years of our lives – the world was a snare and we fell into the vortex till seven children were given to us. We were now removed to Sickle Mill. There worldly prosperity did not attend us, and the transition

from war to peace was a heavy lug upon us; for from a large stock both in the paper & farming line, the decrease in its value, caused a serious loss; yet we continued at Sickle Mill till our family was grown up. I found it necessary to sell my land and it happened at a time when that was also at a very low ebb, & after a few years to give up my business to my son James and retired to the cottage in which I now live. And here by the mercy of God I have been comfortably settled on a limited income.

> This was James III's birthday and as he was married on 28 November 1805 he had spent 54 and not 53 years with his wife Charlotte.

SEPTEMBER

13. [Thursday] Walked to the New Mill and the Orchard. The Mill was sold by auction at the White Horse Inn by F & C Mellersh, and bought by James for 310£ the Directors of the Railroad having previously made it a freehold.

> On 19 May 1857 Simmons had transferred New Mill to James IV, who sold it to the Portsmouth Direct Railway Company for £820 on 23 October 1858.[8] The repurchase price of £310 is clearly a bargain compared with this and also with the £85 per year rent the Appletons agreed to pay on 23 October 1854.

1862

SEPTEMBER

6. [Saturday] Went with Ann to St Mary Cray to see the Mills, neither Mr Joynston nor his son were at home took a note with us written by Mr Cooper and the Manager sent a man to show us over the whole of it. Paper making there is to be seen in perfection. When the paper leaves the wire & comes to the cutting machine finished it is nearly a mile long – steam dried first, then sized & air dried, by fans in cylinders. 300 women in the rag house, the linen & cotton rags separated & every thing in complete order – a place for everything and every thing in its place – about forty tons made weekly and 700 hands employed. Making of envelopes had not began but the machinery ready.

> Simmons and his youngest daughter Ann, who was now living with him, were visiting his daughter Catherine Cooper at Bickley, then in Kent but now in London. William Joynson's paper mill at neighbouring St Mary Cray was probably at this time the largest in the country.[26]

1866

JANUARY

18. [Thursday] James to Farnham on his return he saw a fire in the direction of Barford on his arrival, he found it to be Barford Lower Mill, lately purchased by Mr [blank] and put in repair for his use in making Flocks. The Mill is quite burnt, the house was saved.

> George Roe Warren and Andrew Warren, whose father William had died in 1861, had been using Barford Lower Mill as a half stuff mill up to 1865. Sale particulars of 1891 describe it as a flock mill; the business being carried on by Mr Verstage and formerly by Mr E Rippen.[24]

1868

JANUARY

16. [Thursday] Rain showry weather.

> This is the last entry in the diaries. James Simmons III died three months later and was buried at St Stephen's Church, Shottermill.
>
> The memorial tablet inside the church reads:

IN
MEMORY OF
JAMES SIMMONS,
WHO DIED APRIL 11TH 1868,
AGED 84.
ALSO CHARLOTTE,
HIS BELOVED AND DEVOTED WIFE,
WHO DIED MAY 19TH 1860;
AGED 81.
THEIR GREAT AIM THROUGH LIFE WAS FAITHFULLY
TO SERVE THEIR GOD AND SAVIOUR,
TO ADVANCE THE SPIRITUAL AND TEMPORAL
WELFARE OF THOSE AROUND THEM,
TO PROMOTE THE CAUSE OF EDUCATION
IN THE NEIGHBOURHOOD; AND PERSONALLY ASSIST
IN THE RELIGIOUS TRAINING AND INSTRUCTION
OF THE CHILDREN.
WITH THE KIND ASSISTANCE OF NEIGHBOURS
AND FRIENDS, THEY WERE MAINLY INSTRUMENTAL
IN THE ERECTION OF THIS CHURCH, ALSO
IN BUILDING AND ESTABLISHING SCHOOLS.
"YEA SAITH THE SPIRIT, THAT THEY MAY REST FROM THEIR LABOURS, AND THEIR WORKS DO FOLLOW THEM." REV. 14.13.

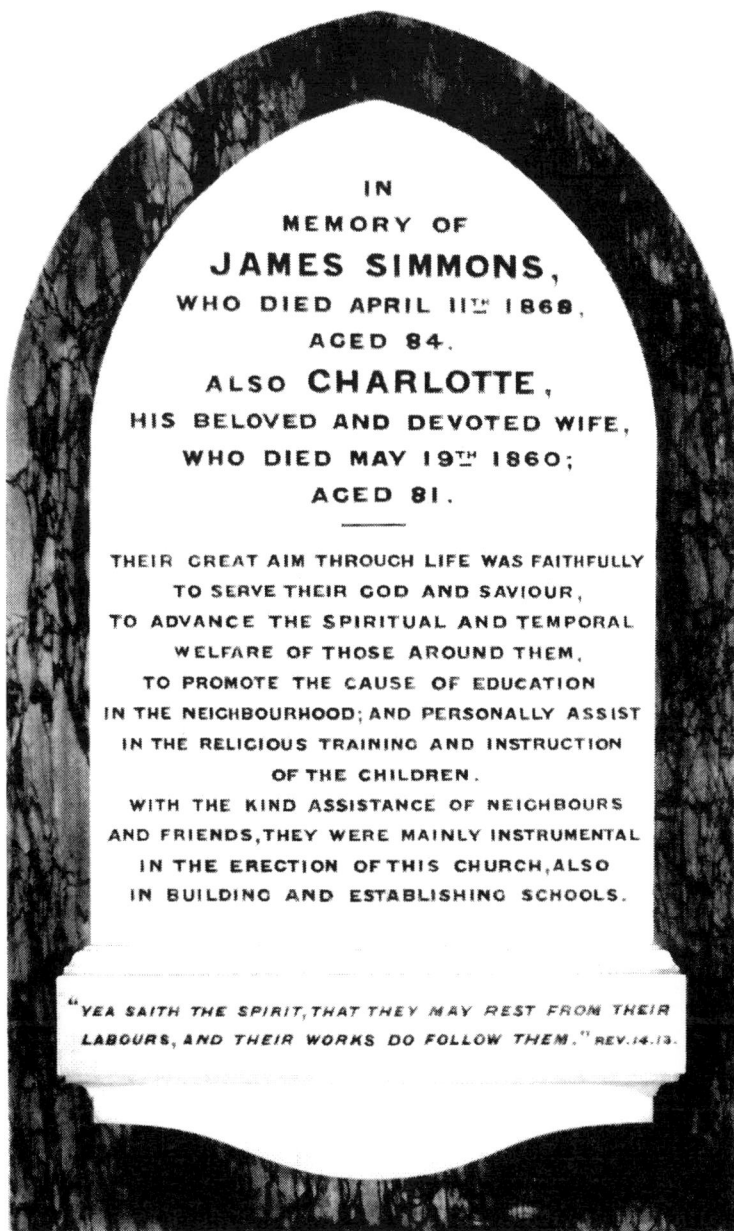

IN
MEMORY OF
JAMES SIMMONS,
WHO DIED APRIL 11TH 1868,
AGED 84.
ALSO **CHARLOTTE,**
HIS BELOVED AND DEVOTED WIFE,
WHO DIED MAY 19TH 1860;
AGED 81.

———

THEIR GREAT AIM THROUGH LIFE WAS FAITHFULLY
TO SERVE THEIR GOD AND SAVIOUR,
TO ADVANCE THE SPIRITUAL AND TEMPORAL
WELFARE OF THOSE AROUND THEM,
TO PROMOTE THE CAUSE OF EDUCATION
IN THE NEIGHBOURHOOD; AND PERSONALLY ASSIST
IN THE RELIGIOUS TRAINING AND INSTRUCTION
OF THE CHILDREN.
WITH THE KIND ASSISTANCE OF NEIGHBOURS
AND FRIENDS, THEY WERE MAINLY INSTRUMENTAL
IN THE ERECTION OF THIS CHURCH, ALSO
IN BUILDING AND ESTABLISHING SCHOOLS.

"YEA SAITH THE SPIRIT, THAT THEY MAY REST FROM THEIR
LABOURS, AND THEIR WORKS DO FOLLOW THEM." REV.14.13.

Figure 18. The memorial tablet inside St Stephen's Church, Shottermill.
(Courtesy Jan Spencer)

Figure 19. Photograph of Sickle Mill House at the left
and part of the mill buildings, which have been converted
into residences, at the right, taken in March 2015.
The stump of an 1854 steam engine shaft behind the
house is featured in figure 20.
(Courtesy Glenys Crocker)

Figure 20. Photograph of the steam engine house
located behind Sickle Mill House, taken in March 2015.
The diaries record that the engine was started on 18 February 1854.
Only the stump of the chimney shaft, which is
shown as a blurred image in figure12, survives.
(Courtesy Glenys Crocker)

Figure 21. Photograph of the decaying New Mill shortly before it was demolished in 1976. The structure is very similar to that shown in the Penfold drawing of 1850 illustrated in figure 14.
(Courtesy Haslemere Educational Museum)

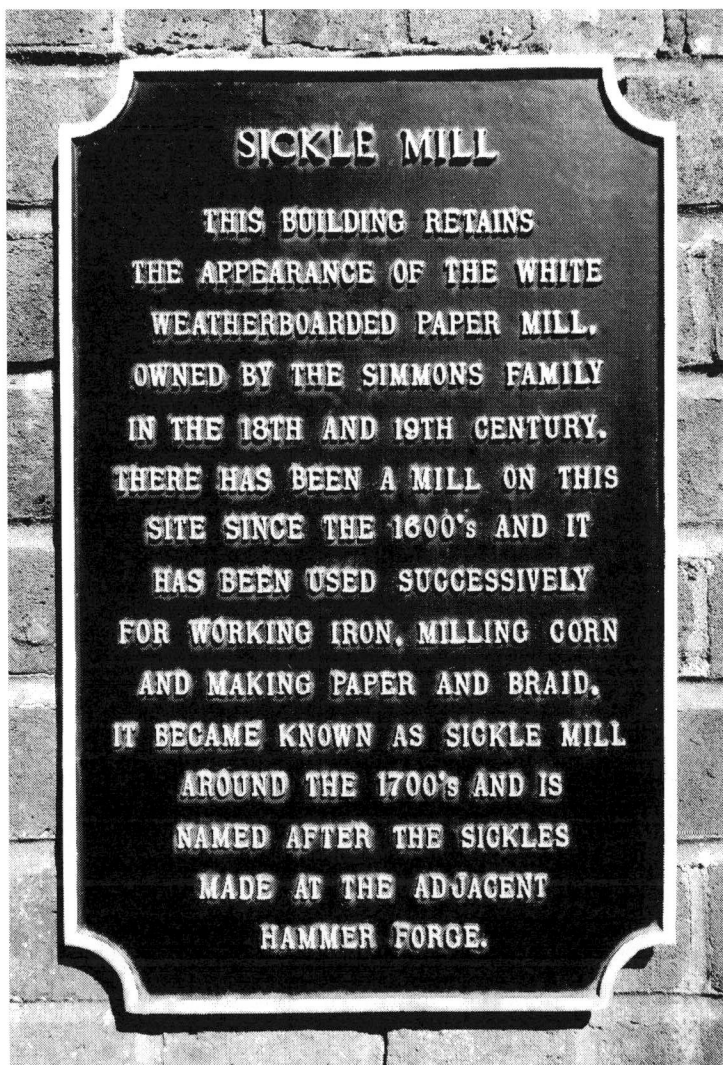

SICKLE MILL

THIS BUILDING RETAINS
THE APPEARANCE OF THE WHITE
WEATHERBOARDED PAPER MILL,
OWNED BY THE SIMMONS FAMILY
IN THE 18TH AND 19TH CENTURY.
THERE HAS BEEN A MILL ON THIS
SITE SINCE THE 1600's AND IT
HAS BEEN USED SUCCESSIVELY
FOR WORKING IRON, MILLING CORN
AND MAKING PAPER AND BRAID.
IT BECAME KNOWN AS SICKLE MILL
AROUND THE 1700's AND IS
NAMED AFTER THE SICKLES
MADE AT THE ADJACENT
HAMMER FORGE.

Figure 22. Photograph of the Sickle Mill Plaque,
taken in April 2015.
(Courtesy Jan Spencer)

Postscript

Haslemere Urban District Council acquired Sickle Mill, which is a Grade II listed structure, in 1930 and it subsequently became part of Waverley Borough Council's property portfolio.[a] In June 1989 it was announced that the Council, had plans for a new high density housing development for the site.[b] However in August 1990 a detailed planning application was prepared to restore and convert the buildings into offices, to build new two-storey offices on adjacent land and to construct nine industrial units on the largely filled mill-pond area behind the mill.[c] This was only two months before the launch of the first edition of this book at a function held in Haslemere Educational Museum. Christine Laver, a paper historian and manufacturer of hand-made paper for book and document repair, attended this launch. She was interested in transferring her business from Somerset to Sickle Mill and, at the suggestion of Ted Orchard who owned two of the Simmons diaries, wrote to the Council with her proposals. It was also arranged for the Wind and Watermills Section of the Society for the Protection of Ancient Buildings to write to the Council about the future of the site. As a result of the strong local and national support for finding the most appropriate use for the buildings and surrounding land, redevelopment was delayed.

In December 1990 and February 1991 Waverley Borough Council obtained new detailed plans of the Sickle Mill building.[d] Then in July 1993 they obtained information on remedial work proposed for the part of Sickle Mill House known as No. 2.[e] However, no further action appears to have occurred until 1996. Unfortunately this was too late for Christine Laver but the new proposals that were then prepared were much more sympathetic to the history of the site.[f] In particular, in June 1996 a planning application for two semi-detached houses in Sickle Mill House and six self-contained apartments in Sickle Mill was received from Thames Valley Housing who had acquired the site from Waverley Borough Council and this proposal was accepted.

In May 1997 it was reported that a badly corroded and damaged historic boiler 13 feet long and 3 feet in diameter (4.0 by 0.9m) had been discovered in a trench being dug for new drains at the Sickle Mill site.[g] This must have been the boiler that James Simmons mentions installing in December 1840. It had been located near a tall chimney that does not survive, but is shown in front of the mill in figure 12. The boiler was of the first type to have a central flue; in 1999 was moved to the Rural Life Centre at Tilford, about 11km north of Sickle Mill, where it was on display for several years.[h] Unfortunately, however, its condition continued to deteriorate rapidly and it has had to be scrapped.

The Thames Valley Housing proposals were accepted and the accommodation was officially opened by the then Mayor of Waverley, Councillor Michael Biddiscombe, in November 1997. The Herons Leisure Centre was built on the site of part of the largely filled mill-pond in 1998, another part having already been developed as Kings Road Industrial Estate. The photograph of Sickle Mill House and part of Sickle Mill, shown in figure 19, and the photograph of the steam engine house located behind Sickle Mill House, shown in figure 20, were taken in 2015. Finally, nothing remains of New Mill; it was demolished shortly after the photograph shown in figure 21 was taken. A plaque summarising the history of the site was attached to the boundary wall of the mill apartments in 1997. A 2015 photograph of the plaque is shown in figure 22.

References

a The Waverley Magazine, February 1998.
b Article headed 'High density housing planned at mill' reported on 16 June 1989 on pages 1 and 2 of a local newspaper, considered to be the *Haslemere Herald*.
c Plans prepared by Modern Design Group Ltd, Architects and Development Consultants.
d Plans of Sickle Mill prepared by Calder Ashby, Chartered Building Surveyors of Chichester, for Waverley Borough Council.
e Details of remedial work prepared by Cooper and Withycombe of Craneigh in July 1993.
f Plans prepared by Delta Chartered Architects of London for Thames Valley Housing.
g *Haslemere Herald*, 23 May 1997, page 2.
h *Haslemere Herald*, 22 January 1999.

Appendix 1
Selective Simmons Pedigree

Abbreviations:
b. born; bapt. baptised; bur. buried; d. died; m. married.
(a) - (d) refer to notes at the end of this appendix.

James Simmons I, son of Humphrey Simmons, was apprenticed at Hurcott paper mill near Kidderminster in 1715, d. 21 Feb. 1777 aged 76, m. Catherine Penfold, 4 Mar. 1735, d. 31 May 1788 aged 79.

They had the following children:

1 John, bapt. Dec. 1736, bur. 10 Nov. 1740.
2 James II, bapt. 2 Apr. 1738, d. 10 Jan. 1790.
3. Thomas, bapt. 3-5 Dec. 1739, bur. 15 May 1743.
4. Humphrey, d. 6 Apr 1817 aged 75, m. 28 Jun. 1772, Ann Rigge, d. 19 July 1812 aged 61.
5 Ann, bapt. 2 May 1744, d. 10 Aug. 1810, m. 26 Apr. 1787 Thomas Penfold, d. 24 Apr. 1827 aged 79 (a).
6 Eliza, bapt. 19 Feb. 1745, d. 4 Dec. 1823, m. 11 May 1766 Richard Stedman, bur. 1 Mar. 1803 aged 63.
7 William, bapt. 1 Jan. 1748, d. 12 Jan. 1801, m. 7 Dec. 1780 Hannah Philps, d. 24 Apr. 1842 aged 90.
8 Joyce, bapt. 10 Oct. 1749, d. 30 Nov. 1822.

William and Hannah (7 above) had the following children:

1 William, b. 22 Dec. 1781, bur. 8 Sept. 1787.
2 James III, b. 2 June 1783, d. 11 Apr. 1868, m. 28 Nov. 1805 Charlotte Newman, b. 4 July 1778, d. 19 May 1860 (b).
3 Catherine, b. 5 Jan. 1785, d. 9 Aug. 1863, m. 22 May 1805 Anthony Newman, bapt. 25 Nov. 1771, bur. 13 Feb. 1829 (b).

4   Elizabeth, b. 15 Dec. 1786, d. 2 Dec. 1852.
5   Ann, b. 5 Dec. 1788, d. 24 Oct. 1860, m. 21 Feb. 1812
    Thomas Newman, d. 23 May 1836 (b).
6   Mary, b. 17 Nov. 1790, d. 17 Apr. 1872, m. 30 Jan. 1828
    John Wornham Penfold, bapt. 3 Feb. 1789, d. 11 Dec. 1873 (c).
7   Sarah, b. 29 Apr. 1798, d. 15 July 1889.

James III and Charlotte (2 above) had the following children:

1   Charlotte Hannah, b. 14 Feb. 1810, d. 1 June 1881, m. 2 Sept. 1856
    John Small (widower), d. 23 Apr. 1887 aged 76.
2   Ellen, b. 7 Sept. 1811, bur. 10 May 1813.
3   Harriot Ann, b. 17 Jan. 1813, bur. 7 May 1816.
4   Catherine, b. 16 May 1814., d. June 1892, m. 14 July 1847
    William Cooper (widower), bapt. 24 June 1806, d. 30 Apr. 1865 (d).
5   James IV, b. 27 Sept. 1815, d. 16 Mar. 1903, m. 16 Oct. 1845
    Ann Lunnon, d. 26 Jan. 1890 aged 67.
6   Ann, b. 15 Apr. 1817, d. 27 Nov. 1884.
7   William, b. 4 Aug. 1820, d. 16 Sept. 1896, m. 24 Apr 1879
    Susannah Carter (widow), d. Jan. 1 1907 aged 90.

James IV and William (5 and 7 above) had no children.

Notes:
a. A family tree of the Penfold family has been published.[6] Thomas
    was related distantly to Ann Simmons's mother Catherine.
b. Anthony and Thomas Newman were the brothers of Charlotte
    Simmons.
c. John Wornham was the son of Thomas and Ann Penfold and
    therefore the first cousin of Mary Simmons.[6]
d. Great-grandparents of Mr. W.J.D. Cooper, who has made the
    diaries available.

Further information about members of the Simmons family is provided
in the Index of Personal Names, and portraits of James III, his wife
Charlotte, his sister Mary and her husband, and his two sons James IV
and William are reproduced in the frontispiece and in figures 7, 15 16,
23 and 24 respectively.

Uncle James' Simmons

MAULL & FOX.   187ᴬ PICCADILLY .W.

Figure 23. Photograph of James Simmons's son James IV
by Maull & Fox of Piccadilly, London.
James IV, who rose to prominence in Local Government and
served as a Justice of the Peace, was the diarist's favourite subject.
(Courtesy W.J.D. Cooper)

Jacques Moll

CHATHAM
AND
SITTINGBOURNE

Figure 24. Photograph of James Simmons's son, William
by Jacques Moll of Chatham and Sittingbourne.
William, who was deaf, lived at Field End
after the death of his father.
(Courtesy W.J.D. Cooper)

Appendix 2

Chronology of the Paper Mills

The abbreviations SFIP and REFIP stand for Sun and Royal Exchange Fire Insurance Policies respectively and pm(s) for either paper-maker(s) or paper mill(s).

1736   James Simmons I leases corn mill in Frensham near Haslemere.[53]

1736   James Simmons I pm. of Frensham insures houses (SFIP 70542).[4]

1736-78 Numerous leases etc. of Simmons estate in Frensham.[53]

1741   William Stanaway and Thomas Howard pms. buried at Haslemere.[4]

1750   William Roe, pm, involved in a riot in Haslemere.[67]

1754   Thomas Kingett described in the Haslemere Poll List as a papermaker.[67]

1758   John Eede apprenticed to James Simmons I pm. of Chiddingfold.[4]

1760-71 Rich. Withall pm. of Haslemere has five children by two wives.[7]

1768   First mention of 'Sickles Paper Mill' on Rocque's map.

1769-70 Richard Withall pm. of Haslemere and Frensham in prison.[5]

1769   J. Simmons I pm. insures Sickle Mill, house etc. (SFIP 277478).[4]

1777   James Simmons I of Frensham paper-maker dies aged 76.[7]

1777   Gilbert White: floods ruin paper mill near Haslemere.[54]

1780   John Tribe apprenticed to William Simmons paper-maker.[4]

1781   Wm. Simmons obtains land on which New Mill was later built.[55]

1782   Thomas Chandler & John Tuckey apprenticed to J. & Wm. Simmons.[4]

1783   J. and Wm. Simmons pms. insure mills, house etc. (SFIP 476691).[5]

1789   John Tilbury & Thos. Eede apprenticed to James & Wm. Simmons.[4]

1790   Daniel Appleton, Pitfold pm, buried at Frensham.[67]

1790   James Simmons II paper-maker of Frensham dies aged 51.[7]

1791c William Simmons owns two paper mills near Haslemere.[4]

1793   Thomas Puttock, a poor child of the parish apprenticed to William Simmons.[66]

1795   William Simmons recorded as owner of Pitfold Mill.[8]

1800   Thomas Tilbury & William Voller apprenticed to William Simmons.[4]

1801   William Simmons paper-maker of Frensham dies aged 53.[7]

1801   Letter from Haslemere causes journeymen to strike at Iping.[13]

1802   J. Simmons III at Sickle, Pitfold & New Mills (REFIP 185092).[5]

1803   Simmons insures 3 pms. occupied by John Howard (REFIP 193979).[5]

1802-3 Land tax. John Howard occupies mills of Wm. Simmons deceased.[56]

1804   'JOHN HOWARD 1804' watermark with Britannia.[29]

1804-11 Land tax. John Howard occupies mills of James Simmons.[56]

1805   Document with 'JOHN HOWARD SURRY' and Britannia watermark.[29]

1812   'J SIMMONS' mould with Post Horn watermark survives.[27]

1812-31 Land tax. James Simmons III owner and occupier of the mills.[56]

1812-41 Many 'J SIMMONS' watermarks with Britannia, Royal Arms etc.[29]

1814c Royal Arms and 'SUPERFINE FINE, JAMES SIMMONS' on wrapper.[23]

1816   Sickle & Shotter (New) mills allocated excise nos. 118 & 480.[5]

1822, 24 Excise change? James Simmons to Richard Smith at 118.[5]

1831   Samuel Philip Everitt is tenant pm. at Pitfold Mill.[1]

1832   James Simmons III tries to sell Pitfold Mill and New Mill.[15]

1832   Excise check. Still James Simmons III pm. at Sickle, no. 118.[5]

1835   James Simmons III uses Pitfold as a half-stuff mill.[1]

1835   Henry Appleton rents part of Pitfold Mill to make braid.[1]

1839   Jos. Puttick & Chas. Harding apprenticed to James Simmons III.[1]

1840   Simmons buys a used paper-making machine for Sickle Mill.[1]

1840   New Mill now used for half-stuff with a greater head of water.[1]

1841-6 Tithe maps. James Simmons holds three paper mills and land.[57, 58, 59]

1847 Excise change. Now James Simmons IV pm. at Sickle Mill, 118.[5]

1847 James Simmons fined £2 for being in breach of excise duties.[60]

1849 John Lill Lightfoot rents Sickle Mill from James Simmons IV.[1]

1849-51 John Wornham Penfold drawings of Sickle Mill and New Mill.[27]

1850 Excise change. Now John Lill Lightfoot pm. at Sickle, 118.[5]

1850 John Lill Lightfoot reported bankrupt.[46]

1851 Excise change. Joseph Fourdrinier tenant pm. at Sickle Mill.[5]

1851 Sickle Mill has one beating engine in use, one silent.[5]

1852 Fourdrinier erects steam engine at Sickle & rents Pitfold.[1]

1852c Fourdrinier's daughter describes tissue making at Sickle Mill.[43]

1852 Fourdrinier fails and Simmons starts making tissue paper.[1]

1853-4 James Simmons IV installs a new steam engine at Sickle Mill.[1]

1853-7 Frensham Common Inclosure Award. Jas. Simmons at Pitfold Mill.[61]

1854 Thomas and William Appleton buy Sickle Mill and rent New Mill.[1]

1857 James III surrenders ownership of New Mill to James IV.[1]

1858 James Simmons IV sells New Mill to the Portsmouth Railway Co.[8]

1860 Tithe alteration. James IV holds Sickle & Pitfold mills.[57, 58]

1860 James IV purchases New Mill from the Railway Company.[1]

1860-70 Appleton Bros., Sickle Mill, Tissues, Copyings, Printings.[62]

1861 Penfold Survey. Henry & Thomas Appleton at Sickle Mill.[63]

1862 New Mill not included in Paper Mills Directory. Closed.[62]

1866-74 Henry Appleton at Pitfold Mill in Directories.[5]

1868 James Simmons III dies aged 84.[8]

1871 Sickle Mill not included in Paper Mills Directory.[62]

Appendix 3

Structure of the Diaries

For each diary the period covered is followed by the number of sections, the number of pages and the height, width and thickness of the booklet. The type of paper used is then specified together with its colour and thickness and, in the case of laid paper, the spacing of the chain and laid lines. Any watermarks present are then described, the abbreviations FL, GR, PH and RA being used for Fleur-de-Lis, 'GR' Royal cypher, Post Horn and Royal Arms designs respectively. Many of these watermarks are illustrated in figures 25, 26 and 27. Finally the cover, if there is one, is described. All measurements are given in millimetres (mm) where 1 inch equals 25.4mm. Note that a section is a gathering of folded sheets, a sheet is folded once to make two leaves, and a leaf has two sides and hence gives two pages.

1. 16 Aug. - 22 Oct. 1831. One section. 88 pages. 184 x 117 x 5.4mm.
   Laid white 0.12mm thick. Chains 23.0mm, lines 1.05mm.
       'JS' over '1830'.
   No cover.

10. 14 Dec. 1834 - 30 June 1835. One section. 112 pages. 191 x 120 x 6.5mm.
    Sheet 1. Wove cream 0.11mm thick.
    Sheets 2-28. Wove cream 0.10mm thick.
        'SIMMONS' over '1833'.
    Cover of wove buff coarser paper 0.27mm thick

11. 2 July - 5 Nov. 1835. One section. 64 pages. 185 x 117 x 4.0mm.
    Wove cream 0.11mm thick.
        'SIMMONS' over '1833'.
    Cover of wove buff coarser paper glued to two outer pages, 0.34mm thick.

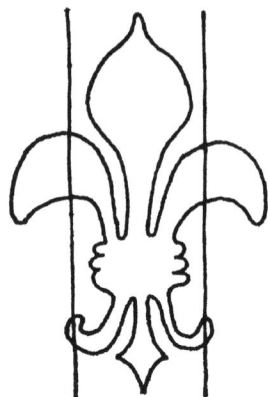

Figure 25. Royal Arms with GR, Fleur-de-Lis, script JS with 1837
and JS with 1830 watermarks (x 0.85) traced from Simmons
paper used in diaries 49, 15, 51 and 1 respectively.

The paper containing the Royal Arms with the GR cypher was probably
made by James I or one of his sons but the remaining examples are in
paper made by James III. All designs, except the 1837 countermark, are in
laid paper, the positions of the chain lines being indicated. The chain and
laid line spacings are given in this appendix.

12. 5 Nov. 1835 - 23 Apr. 1836. One section. 72 pages. 185 x 119 x 4.1mm. Wove cream 0.10mm thick.

     'SIMMONS' over '1833'.

   Cover of wove thin grey-buff paper glued to two outer pages, 0.23mm thick.

13. 24 Apr. - 13 Aug. 1836. One section. 64 pages. 187 x 117 x 3.6mm. Wove cream 0.10mm thick.

     'SIMMONS' over '1833'.

   Cover of wove pink paper glued to two outer pages, 0.33mm thick.

14. Aug. - 24 Dec. 1836. One section. 64 pages. 185 x 116 x 3.7mm. Wove cream 0.10mm thick.

     'SIMMONS' over '1833'.

   Cover of wove pink coarser paper glued to two outer pages, 0.25mm thick.

     '1822'

15. 25 Dec. 1836 - 25 Aug. 1837. One section. 124 pages. 190 x 124 x 6.8mm.

   Sheets 1-4. Laid blue-white 0.09mm thick. Chains 25.5mm, lines 1.0mm.

     'SHEPHEARD & SUTTON' over '1835' with FL.

   Sheets 5-8. Laid blue-white 0.10mm thick. Chains 25.0mm, lines 1.0mm.

     'JS' over '1833' with FL.

   Sheets 9-20. Laid blue-white 0.09mm thick. Chains 24.5mm, lines 1.0mm.

     'JS' over '1835' with FL.

   Sheets 21-24. Laid blue-white 0.14mm thick. Chains 23.5mm, lines 1.0mm.

     'JS' over '1833' with PH on shield over script 'JS'.

   Sheets 25-31. Wove cream 0.11mm thick.

     'SIMMONS' over '1833'.

   Cover of wove cream coarser paper glued to outer pages, 0.26mm thick.

Figure 26. Ionian with UIS and Fleur-de-Lis with JS watermarks (x 0.74) traced from Simmons paper used in diaries 23 and 21 respectively. These examples are in paper made by James Simmons III. The designs are in laid paper, the positions of the chain lines being indicated. The chain and laid line spacings are given in this appendix.

16. 27 Aug. 1837 - 23 Feb. 1838. One section. 96 pages. 197 x 121 x 3.8mm.
    Laid cream 0.08mm thick. Chains 25.0mm, lines 0.90mm.
    'JS' over '1837' with FL.
    No cover.

17. 23 Feb. - 10 Oct. 1838. One section. 124 pages. 191 x 125 x 6.1mm.
    Sheets 1-4, 13-24, 27-31. Laid cream 0.09mm thick. Chains 24.5mm, lines 1.0mm.
    'JS' over '1837' with FL.
    Sheets 5-12. Laid cream 0.11mm thick. Chains 24.0mm, lines 1.05mm.
    'J Simmons' over '1835' with PH on shield over script 'JS'.
    Sheets 25-26. Laid blue-white 0.09mm thick. Chains 23.5mm, lines 1.10mm.
    'CHARLES SKIPPER & EAST'.
    No cover.

18. 11 Oct. 1838 - 12 Oct. 39. One section. 192 pages. 191 x 125 x 8.3mm.
    Laid blue-white 0.09mm thick. Chains 24.5mm, lines 1.05mm.
    'JS' over '1839' with FL.
    No cover.

19. 13 Oct. 1839 - 22 Mar. 1810. One section. 92 pages. 185 x 122 x 5.3mm.
    Sheets 1-2. Laid cream 0.10mm thick. Chains 24.5mm, lines 1.05mm. FL.
    Sheets 3-6. Laid cream 0.09mm thick. Chains 23.5mm, lines 1.10mm.
    'JS' over '1838' with FL.
    Sheets 7-10, 17-19. Laid cream 0.13mm thick. Chains 23.5mm, lines 1.35mm.
    'J SIMMONS' over '1839' with FL on shield over script 'JS'.
    Sheets 11-12. Laid cream 0.12mm thick. Chains 23.5mm, lines 0.95mm.
    'JS' over '1839'.
    Sheets 13-16. Laid cream 0.10mm thick. Chains 24.5mm, lines 1.00mm.
    'JS' over '1838' with FL.

Figure 27. Post Horn with JS and Britannia with J SIMMONS 1841
watermarks (x 0.67) traced from laid paper made by James Simmons III
and used in diaries 19 and 22 respectively.
The designs are in laid paper, the positions of the chain lines beng
indicated. The chain and laid line spacings are given in this appendix.

Sheets 20-23. Laid cream 0.12mm thick. Chains 24.0mm, lines 1.00mm.

'JS' over '1837' with PH on shield over script 'JS'.

No cover.

20. 23 Mar. 1840 - 14 Mar. 1841. One section. 160 pages. 193 x 125 x 7.6mm.

Sheets 1-8. Laid cream 0.10mm thick. Chains 24.5mm, lines 1.05mm.

'JS' over '1835' with FL.

Sheets 9-32, 37-40. Laid cream 0.09mm thick. Chains 24.5mm, lines 1.00mm.

'JS' over '1838' with FL .

Sheets 33-36. Laid cream 0.09mm thick. Chains 24.4mm, lines 0.95mm.

'JS' over '1836' with FL.

Cover of wove coarse buff paper, 0.15mm thick.

'1830'.

21. 20 Mar. - 2 Sept. 1841. One section. 88 pages. 186 x 125 x 6.7mm.

Laid cream 0.15mm thick. Chains 27.5mm, lines 1.35mm.

'J SIMMONS' over '1839' with FL on shield over script 'JS'.

22. 3 Sept. 1841 - 31 Mar. 1842. One section. 96 pages. 197 x 127 x 4.7mm.

Laid cream 0.09mm or 0.12mm thick. Chains 26.0mm, lines 1.10mm vertical.

Two variants of 'J Simmons' over '1841' with Britannia.

No cover.

23. 3 Mar. - 18 Sept. 1842. One section. 96 pages. 197 x 127 x 6.5mm.

Laid cream 0.14mm thick. Chains 25.5mm, lines 1.0mm vertical.

Two or more variants of 'UIS' with Ionian Coat of Arms.

No cover.

24. 19 Oct. 1842 - 12 May 1843. One section. 120 pages. 203 x 133 x 7.7mm.

Wove smooth cream 0.1 2mm thick.

Script 'JS' over '1840'.

Cover of wove buff coarser paper glued to outer pages, 0.33mm thick.

25. 25 May 1843 - 11 Feb. 1844. One section. 110 pages. 194 x 130 x 6.8mm.
Wove smooth cream 0.11mm thick.
    Script 'JS' over '1840'.
Cover of wove buff coarser paper glued to outer pages, 0.38mm thick.

26. 12 Feb. - 15 Dec. 1844. One section. 116 pages. 194 x 131 x 7.1mm.
Wove smooth cream 0.11mm thick.
    Script 'JS' over '1840'.
Cover of wove buff coarser paper glued to outer pages, 0.37mm thick.

27. 21 Dec. 1844 - 4 Aug. 1845. One section. 92 pages. 203 x 125 x 3.6mm.
Wove smooth cream 0.08mm thick.
    Script 'JS' over '1840'.
No cover.

28. 8 Aug. 1845 - 24 Mar. 1846. One section. 116 pages. 195 x 118 x 10.9mm.
Wove smooth cream 0.18mm thick.
Cover laid very coarse grey-brown 0.33mm thick. Chains 25.5mm, lines 1.0mm.
    '------ULORY'?

31. 25 Apr. - 26 Oct. 1847. One section. 92 pages. 212 x 142 x 6.6mm.
Wove smooth cream 0.14mm thick.
No cover.

32. 3 Oct. 1847 - 22 Apr. 1848. One section. 96 pages. 216 x 142 x 6.7mm.
Wove smooth cream 0.14mm thick.
No cover.

33. 23 Apr. - 19 Nov. 1848. One section. 124 pages. 216 x 143 x 6.6mm.
Wove smooth cream 0.14mm thick.
No cover.

35. 30 Apr. - 31 Dec. 1849. One section. 124 pages. 216 x 143 x 10.1mm.
Wove smooth cream 0.16mm thick.
No cover.

36. 1 Jan. - 31 Dec. 1850. Two sections. 140pages. 215 x 147 x 10.5mm.
     Section 1. 29 sheets. Wove smooth cream 0.14mm thick.
     Section 2. 6 sheets. Wove smooth cream 0.14mm thick.
     No cover.

37. 1 Jan. - 31 Dec. 1851. One section. 128 pages. 216 x 149 x 8.3mm.
     Wove smooth cream 0.13mm thick.
     No cover.

38. 1 Jan. - 31 Dec. 1852. One section. 144 pages. 216 x 154 x 11.8mm.
     Wove smooth cream 0.15mm thick.
     Two cover sheets of wove rough pink paper each 0.25mm thick.
     223 x 160mm.

39. 1 Jan. - 31 Dec. 1853. Two sections. 152 pages. 215 x 150 x 11.2mm.
     Section 1. 36 sheets. Wove smooth cream 0.15mm thick.
     Section 2. 2 sheets. Wove smooth cream 0.15mm thick. 210 x
     143mm.
     No cover.

40. 1 Jan. - 31 Dec. 1854. One section. 152 pages. 213 x 150 x 10.0mm.
     Wove smooth cream 0.13mm thick.
     No cover.

43. 1 Jan. - 15 Dec. 1857. Two sections. 120 pages. 235 x 160 x 8.8mm.
     Section 1. 18 sheets. Wove cream 0.16mm thick.
         'JS' over '1837'.
     Section 2. 12 sheets. 242 x 150mm. Wove smooth cream 0.13mm
     thick.
     No cover.

44. 1 Jan. 1858 - 31 Dec. 1859. Eleven sections. 250 pages. 243 x 152
     x 17.5mm.
     Sections 1-7. 8, 8, 7½, 4, 8, 5, 6 sheets. Wove smooth cream
     0.11mm thick.
     Sections 8-11. Each 4 sheets. Wove cream 0.21mm thick.
         'SIMMONS' over '1817'.
     Front cover formed from first two pages glued together, 0.31mm thick.

45. 1 Jan. - 31 Dec. 1860. Five sections. 164 pages. 226 x 142 x 11.4mm. Sections 1-5. 3, 8, 12, 9, 9 sheets. Wove mainly smooth cream 0.11mm to 0.18mm thick.
Front cover formed by gluing extra wove sheet to front page, 0.32mm thick.

46. 1 Jan. 1861 - 25 Jan. 1862. Nine sections. 140 pages. 227 x 143 x 7.3mm.
Sections 1-8. Each 4 sheets. Wove cream 0.11mm thick.
Section 9. 3 sheets. 223 x 143mm. Wove grey 0.06mm thick.
Front cover of extra single sheet of wove smooth cream 0.12mm thick.

47. 26 Jan. - 31 Dec. 1862. Eight sections. 112 pages. 200 x 127 x 6.9mm.
Sections 1-2. Each 2 sheets. Wove smooth cream 0.12mm thick.
Sections 3-8. Each 4 sheets. Wove smooth cream 0.12mm thick.
No cover.

48. 1 Jan. - 31 Dec. 1863. Eight sections. 124 pages. 209x 135x 10.2mm.
Sections 1-2. Each 4 sheets. Wove smooth cream 0.20mm thick.
Section 3. 4 sheets. Wove smooth cream 0.10mm thick.
Section 4. 4 sheets. Laid white 0.17mm thick. Chains 27mm, lines 1.20mm.
    'T EDMONDS' over '1832' with FL on shield over script 'TE'.
Section 5. 3 sheets. Wove cream 0.18mm thick.
    'T EDMONDS' over '1832' with 'NOT BLEACHED'.
Section 6. 4 sheets. 202 x 130mm. Wove cream 0.15mm thick.
    'T EDMONDS'.
Section 7. 4 sheets. 193mm x 125mm. Same as section 4 but 0.16mm thick.
Section 8. 4 sheets. 195 x 122mm. Wove cream 0.11mm thick.
    'SIMMONS' over '1830'.
Insert A. 2 sheets. 197 x 125mm. Wove cream 0.13mm thick.
Insert B. 1 sheet. 131 x 1 05mm. Laid blue-white 0.09mm thick. Chains 23mm, lines 0.95mm.
    Part of PH on shield over script 'TE'.
No cover.

49. 1 Jan. - 31 Dec. 1864. Six sections. 132 pages. 207 x 142 x 8.8mm.
     Sections 1-5. 6, 4, 5½, 6, 5 sheets. Laid cream 0.13mm thick.
     Chains 24mm, lines 1.10mm vertical.
         GR with 1714-1801 version of RA.
     Section 6. 6½ sheets. Wove cream 0.09mm thick.
         Script 'JS' over '1840'.
     Cover of wove mid-blue paper glued to outer pages of sections 1
     and 5, 0.34mm thick.

51. 1 Jan. 1866 – 30 July 1867. Four sections. 238 pages. 217 x 135 x
     11.1mm.
     Sections 1-4. 14, 14, 17½, 14 sheets. Wove cream 0.09mm thick.
         Script 'JS' over '1837'.
     Cover laid smooth grey glued to outer pages, 0.33mm thick.
     Chains 26mm, lines 1.1mm.

52. 31 July 1867 - 16 Jan. 1868. One section. 80 pages. 210 x 133 x 3.6mm.
     Sheets 1-6. Laid white 0.10mm thick. Chains 24mm, lines 1.05mm.
         'J SIMMONS' over '1838' with PH on shield over script 'JS'.
     Sheets 7-20. Wove cream 0.09mm thick.
         Script 'JS' over '1840'.
     No cover.

AS   10 June - 31 Oct. 1859. One section. 80 pages. 197 x 131 x 5.0mm.
     Wove cream 0.11mm thick.
         Script 'JS' over '1837'.
     Cover of wove light brown coarse paper glued to outer pages,
     0.39mm thick.

AS = Diary of Ann Simmons, daughter of James III.

Appendix 4
Glossary of Technical Terms

Asp: Prop on the bridge across a vat against which the coucher rests a mould supporting a wet sheet of paper.

Beating Engine: Large tank in which rag fibres mixed with water are cut, pounded and frayed between a rotating roller and a bed-plate, which are fitted with blades or bars to produce stuff. Also known as Hollander.

Bleaching: Removal of colour from disintegrated rags by exposure to the sun, chlorine gas or chloride of lime.

Boiler: Container for boiling rags or other raw materials under pressure in a caustic soda solution.

Breaker: Machine in which rags are broken down into individual fibres, bleached and washed to produce half-stuff.

Broke Paper: Sheets of paper with short tears, small holes, bubbles, wrinkles, drops of size etc.

Caps: Wrapping paper such as that used by grocers.

Chest: See *Stuff Chest.*

Coucher: Workman in a hand-made paper mill who takes the mould with its sheet of wet paper from the vatman, inverts it in order to place the paper on a piece of felted wool fabric and repeats the process to produce a pile or post of usually 144 sheets.

Countermark: Watermark in paper containing the name or initials of the papermaker, the date, and occasionally the name of the mill or county.

Cut: Watercourse which has been straightened.

Cutter: Equipment which cuts machine-made paper into sheets.

Glossary of Technical Terms

Dandy Roll: Light skeletal wire cylinder used with a Fourdrinier machine. It presses on the wet paper and can be used to produce watermarks.

Deckle: Loose frame of wood which the vatman places around his mould to limit the size of the sheet of paper he is making and which he removes before passing the mould to the coucher. Alternatively the straps which restrict the width of paper made on a machine.

Demy: Paper measuring 17½ by 22½ inches (432 x 445mm) for printing and 15½ by 20 inches (394 x 508mm) for writing and drawing. From the French *demi,* half.

Devil: Machine for tearing rags and removing dust and dirt. Also used in the woollen industry. See *Willow.*

Drying Cylinders: Heated highly polished metal cylinders of paper-making machine against which the paper is pressed by felts and dried.

Drying House: Building or loft with opening shutters in which sheets of paper are hung to dry on cowhair ropes stretched across horizontal frames known as tribles which are supported by vertical posts called standers.

Dusting House: Building, sometimes part of rag house, in which dust is removed from rags.

Duty: Tax on paper introduced in 1712 and finally repealed in 1861.

Engine: See *Beating engine.*

Engine House: Building housing the beating engines.

Felts: Pieces of felted woollen fabric, which absorb water from individual sheets of paper, or continuous felted blankets and rollers used for the same purpose with machines.

Finishing Room: See *Salle.*

Foolscap: Paper measuring 13½ by 17 inches (343 x 432mm) for printing and 13½ by 16½ (343 x 419mm) for writing. Named after the watermark of a court jester which was replaced in England by Britannia.

Glazing: Producing a smooth finish on sheets of paper by polishing with a stone, hammering, or pressing and shearing between metal plates. Alternatively passing machine-made paper through rollers.

Gudgeon: Piece of metal let into the end of a wooden shaft.

Half Stuff: Coarse pulp with separated fibres, often bleached, produced in a beating engine.

Hog: Stirring device which prevents the fibres of the stuff settling at the bottom of the vat.

Knotter: Box-like apparatus invented in 1830 with a base made up of a series of slitted brass plates. These filter knotted fibres and foreign substances from the stuff before it runs into the vat.

Laid: Paper with characteristic straight-line markings of the closely spaced wires of a mould and the more widely spaced perpendicular chain markings.

Layer: Workman in a hand-made paper mill who places damp warm sheets of felt ready to receive sheets of paper from the coucher, presses the resulting post of paper and then separates the sheets of paper from the pieces of felt.

London Fines and Seconds: Good and poor quality rags respectively.

Lumber hand: Wrapping paper measuring about 23 by 18 inches (584 x 457mm).

Machine: Paper-making machine which produces a continuous roll of paper, first developed successfully by Bryan Donkin at Bermondsey in 1802-6 with finance from the London wholesale stationers Henry and Sealy Fourdrinier.

Mechanical Press: See *Press.*

Millboard: Thick stiff board used for book covers made by pasting together several sheets of paper.

Mould: Rectangular frame of wood supporting a bronze wire cover which the vatman dips into the vat to form a sheet of paper. See *Laid* and *Wove.*

News: Newsprint or paper made specifically for printing newspapers.

Pit Wheel: Large cog wheel mounted on the shaft of a waterwheel and which turns in a pit inside the mill building.

Plate: Paper suitable for copper-plate printing.

Poshole: Heater used to warm the stuff in a vat. Known as *Pistolet* in France.

Post: Stack of 144 sheets of paper interleaved with felts.

Press: Machine, operated mechanically with a screw or hydraulically, for squeezing water from newly formed wet sheets of paper (vat-press) or for flattening sheets of dry paper (dry-press).

Rag House: Building in which rags for paper-making are dusted, cut to remove seams and sorted.

Rags: Raw material from which until about 1860 almost all paper was made. Rags were sorted into several qualities including: superfine; fine; stitches of fine; middling; seams and stitches of middling; coarse; the very coarse being rejected for white paper but used for brown paper.

Ream: Package of 480 sheets of paper.

Salle: Finishing room where paper is inspected, polished, sorted into sizes and qualities and packed.

Size: Thin glutinous liquid produced from hides, hoofs and bones of animals. Size improves the strength and durability of paper.

Sizing House: Building or room in which a workman, known as a sizer, passes sheets of pressed paper through a bath of warm size to make them less absorbent.

Small Hand: Wrapping paper measuring 16 by 19¾ inches (406 x 502mm). Originally named after the watermark of a hand or glove.

Sol: See *Salle.*

Stuff: Fine pulp ready for making paper.

Stuff Chest: Container in which the stuff is stored and agitated to avoid settling before being transferred to the vat.

Suction Box: Box placed under the wire of a Fourdrinier machine to extract water from the wet paper passing over it.

Tail: Watercourse leading away from a mill.

Team: Wagon and horses.

Thill: Shaft of a cart.

Vat: Large open tank containing warm stuff into which the vatman dips his mould to make a sheet of hand-made paper or from which pulp flows on the endless wire web of a paper-making machine.

Vatman: Skilled workman in a hand-made paper mill who forms a sheet of wet paper by dipping his mould into the pulp in the vat and then shaking it to remove surplus water and cause the fibres to intertwine.

Watermark: Mark on paper arising from a wire design, typically Britannia, a Crown, the Royal Arms, a Fleur-de-Lis, a Lion, 'Pro Patria' or a Horn, sewn with fine wire on the mould or dandy roll. 'Shadow' watermarks, such as the portraits in modern bank note papers, are produced by shaping the wove wire.

Wheel House: Building housing a waterwheel at a mill.

Willow: Machine with revolving cylinders armed with teeth for tearing rags and removing dust and dirt. Also used in the woollen industry. See *Devil.*

Wire: Endless web of wire on which the paper is formed on a paper-making machine.

Wove: Paper with ill-defined markings of the woven fine wires of a mould.

Wrappers: Paper produced specifically for wrapping goods and parcels.

Appendix 5
Making Paper on a Simmons Mould

Two leaves at the end of the first edition of this book were of paper made in 1989 on a James Simmons mould bearing a post horn watermark and the countermark 'J SIMMONS 1812'. This mould, which is remarkably well preserved, was located at Haslemere Educational Museum during a search, prompted by the discovery of the diaries, for other archive material and artefacts associated with the Simmons family. A photograph of part of its laid cover, showing the post horn device, is reproduced as figure 4. In 1812 Simmons had just taken over the Haslemere paper mills and this must have been one of his new moulds. It was made by John Green, jnr., of Maidstone, measures 1030 by 424mm and was used to make two separate sheets of paper.

Permission was obtained from the curator and the trustees of the Museum to attempt to make paper on this historic mould. However, before this could be done, it had to be strengthened as some of the stitches of fine wire fixing the laid cover to the backing ribs are broken. This was achieved by tying the neighbouring chain lines to the ribs using fine nylon fishing line. The nylon is not visible from the front of the mould, is unobtrusive at the back and if desired could easily be removed. Another problem was that the deckle had not survived, but a former joiner agreed to make one from the measurements provided. These included a slight downwards bowing of the long edges so that when the vatman grips together the ends of the mould and deckle, the latter flexes to give a tight fit along all four edges. Also, as the mould was designed to make two sheets of paper, the central dividing rib of the deckle had to be bowed as well. The new deckle, made from pine, appeared to fit the mould well, but when compared with an available historic deckle was found to be about four times as heavy.

Figure 28. Ian Wilcox, the vat man, and Pascal Web, the coucher,
making paper on the 1812 Simmons mould
at Wookey Hole paper mill in 1989.

In the meantime it was discovered that known craft paper-makers did not have vats large enough for the double mould. The paper mill at Wookey Hole near Wells in Somerset, which demonstrates paper-making to tourists, was therefore contacted and they agreed to attempt to make paper using the mould. This seemed appropriate as Simmons mentioned this mill in his diary on 30th January 1843. The mould with its new deckle was taken to the mill on 4th December 1989. The vatman was able to cope with the weight but unfortunately the deckle was too rigid so that when he gripped its ends it was the mould which flexed and not the deckle. This might not have mattered with a new mould, but the frame of the Simmons mould had been repaired at some time, at the mid point of one of its long edges. This edge tended to hinge at the repair, allowing the stuff to flow off the mould under the deckle. The only solution was for the coucher to grip the deckle to the mould at this point, while the vatman was forming the sheets. Clearly this was tedious and tiring but fortunately it proved to be successful. Enough sheets were made to include one in each copy of the first edition of the book. The mould does not appear to have suffered from being used again after 177 years and has been returned to the museum together with its new deckle.

Thanks are due to Diana Smith and Richard Muir of Haslemere Educational Museum for allowing the mould to be used, to Peter Bower for advising on the design of the deckle, to Bernard Oddie for making the deckle and finally to Barney Butter, Ron Grant, Pascal Webb and Ian Wilcox of Wookey Hole paper mill for producing the paper.

References

1   Manuscript Diaries of James Simmons, 1831-68. Surrey History Centre
    1657/1/1 - 1657/3/8; (37 diaries) and Haslemere Educational Museum
    LD.5.515, LD.5.502 (diaries for 1850 and 1851).

2   Hills, R.E., *Papermaking in Britain 1488-1988,* Athlone, London, 1988.

3   Hunter, D., *Papermaking, the History and Technique of an Ancient Craft,* Dover,
    New York, 1978.

4   Shorter, A.H., *Paper Mills and Paper Makers in England 1495-1800,*
    Paper Publications Society, Hilversum, 1957.

5   Simmons, H.E.S., *'The Simmons Water Mills Collection',* unpublished (1940s).
    Science Museum Library, London.

6   Swanton, E.W., ed., *Bygone Haslemere* (expanded edition including
    'Monumental Inscriptions'), West, Newman & Co., London, 1914.

7   *Haslemere Parish Registers, 1573-1812,* Surrey Par. Reg. Soc., vols. 8, 9; 1906.

8   Cooper Archives. Records of the Cooper and Simmons families of Haslemere.
    These papers have now been transferred to the Surrey History Centre, Woking.

9   Bowles, N. and Kane, M., *The Southern Wey; A Guide,* River Wey Trust,
    Liphook, 1988.

10 Map of Surrey by Christopher and John Greenwood, 1823.

11 Paper-maker's Mould used by James Simmons held at Haslemere Educational
    Museum.

12 Spicer, A.D., *The Paper Trade,* Methuen, London, pp. 109-24, 247-51; 1907.

13 Coleman, D.C., *The British Paper Industry, 1495-1860,* Clarendon, Oxford, 1958.

14 Krill, J., *English Artists Paper, Renaissance to Regency,* Trefoil, London, 1987.

15 Sale Particulars of Simmons Estate (1832); including Pitfold and New Mills.
    Farnham Museum 38/J/5.

16 Crocker, A. and Crocker, G., *Catteshall Mill,* Res. Vol. 8, Surrey Archaeol. Soc.,
    Guildford, 1981, pp. 13-6.

17 Shorter, A.H., 'Paper Mills in Sussex', *Industrial Past* 10(1), 1983, pp. 26-9.

18 Pigot, J. & Co., *Royal National & Commercial Directory & Topography,* 1839.

19 Ince, E.E., *Miscellany of the First Baptist Church, Middletown,* 1889.

20 Hillier, J., *Old Surrey Water-Mills,* Skeffington, London, 1951, pp. 71-2, 78-86.

References

21 Evans, J., *The Endless Web, John Dickinson & Co. Ltd., 1804-1954,* Jonathan Cape, London, 1955, pp. 70-86.

22 *Spicer Family Tree,* Private Publication. Copy at Surrey Archaeol. Soc. Library.

23 Shorter, A. H., 'Paper Mills in Hampshire', *Proc. Hants. Field Club & Arch. Soc.,* 18(1), 1952, pp. 1-11.

24 Warren Collection. Records and Artifacts held by Mr J. Warren of Headley, Hants.

25 Crocker, A., *Paper Mills of the Tillingbourne,* Tabard, Oxshott, 1988.

26 Shorter, A.H., *Paper Making in the British Isles,* David & Charles, Newton Abbot, 1971.

27 Archive Material held at Haslemere Educational Museum.

28 White, H.P., *A Regional History of the Railways of Great Britain: 2. Southern England,* Kelley, New York, 1970, pp. 108-34.

29 Crocker, A., 'Watermarks in Surrey Hand-made Paper'. *Surrey History,* 3(1), 1984, pp. 2-16.

30 Crocker, A., 'The Paper Mills of Surrey', *Yearbook Int. Assoc. Paper Historians, 7,* 1988, pp. 10-27.

31 Weinreb, B. and Hibbert, C., *The London Encyclopaedia,* BCA, London, 1985, pp. 483- 4.

32 Boase, F., *Modern English Biography,* Cass, London, 1965.

33 *Dictionary of National Biography,* Oxford University Press, Oxford, since 1917.

34 Cross, E.R., ed., *A Brief Account of Bryan Donkin* FRS *and of the Company he Founded 150 Years Ago,* Bryan Donkin Co., Chesterfield, 1953.

35 Mayes, J.L., 'Paper in the Wye Valley', in *Three Hundred Years in Paper,* Mandl, G.T., ed., Private Publication, 1985, pp. 133-58.

36 Tomlinson, C., ed., *Cyclopaedia of Useful Arts and Manufactures,* George Virtue & Co., London, 1852.

37 *Palmer, A.W., A Dictionary of Modern History 1789- 1945,* Penguin, Harmondsworth, 1962.

38 *Penny Magazine Monthly Supplement,* 96, pp. 377-84, 28 Sept. 1833.

39 Vine, P.A.L., *London's Lost Route to the Sea,* David & Charles, Newton Abbot, 1965.

40 *Sussex Agricultural Express,* 12 June, 1841, p. 1, col. c.

41 Hadfield, C., *Atmospheric Railways, a Victorian Venture in Silent Speed,* David & Charles, Newton Abbot, 1967, pp. 78, 214.

42 Turner, J.T.H., *The London Brighton & South Coast Railway: 1. Origins and Formation,* Batsford, London, 1977, pp. 239-256.

43 Edwards, R., 'Henry Fourdrinier's Descendants', *The Paper Maker & British Paper Trade Journal,* 157, 1969, pp. 32-3.

44 Gateacre, V.E. and Winsor, D., *Wookey Hole,* Wookey Hole Caves Ltd., 1977.

References

45   Watson, N., *The Last Mill on the Esk,* Scottish Academic Press, Edinburgh, 1987, pp. 12-13.

46   *Haslemere Herald,* 9 Aug. 1985.

47   Gilchrist, H.H., *Anne Gilchrist, Her Life and Writings.* Fisher Unwin, London, 1887 pp. 161-65, 171.

48   Austen, B., Cox, D. and Upton, J., *Sussex Industrial Archaeology: A Field Guide,* Phillimore, Chichester, 1985, p. 44.

49   *Sussex Agricultural Express,* 20 Nov. 1852, p. 6, col. f.

50   Frensham Parish Registers. Transcript at Surrey History Centre FREN/1/1; see also www.ancestry.co.uk.

51   Cossons, N., *The BP Book of Industrial Archaeology,* 2nd. ed., David & Charles, Newton Abbot, 1987, pp. 78-80.

52   English, W., *The Textile Industry,* Longmans, London, 1969, pp. 183-93.

53   Bishopric of Winchester Archives. Hampshire Record Office; Eccles II.

54   Johnson, W., ed., *Gilbert White's Journals,* David & Charles, Newton Abbot, 1970, pp. 140-1.

55   Linchmere Court Roll. West Sussex Record Office; Cowdray Manuscripts.

56   Haslemere Land Tax Records. Surrey Record Office; SRO QS6/7.

57   Frensham Tithe Apportionment and Map of 1839 and Alteration of 1860.

58   Haslemere Tithe Apportionment and Map of 1841 and Alteration of 1860.

59   Linchmere Tithe Apportionment and Map of 1846.

60   *'Return of Convictions of Manufacturers of Paper, 1840-8'.* House of Commons, 19 June 1849.

61   Frensham Commons, Farnham Manor Inclosure, 1853-7, Surrey Record Office; SRO QS6/4/55. (Now Surrey History Centre, SHC QS6/4/55).

62   *Paper Mills* Directory, 1860-71.

63   Penfold, J.W., 'Survey of Haslemere, 1861', unpublished. Held at Godalming Museum.

64   Crocker, A., 'Paper excise stamps on a re-used Haslemere ream-wrapper', *Surrey Archaeological Collections,* 83, 1996, pp. 159-64. Reprinted in *The Quarterly,* 40, Oct. 2001.

65   Dagnall, H. *The Taxation of Paper in Great Britain,* A History and Documentation, the author and the British Association of Paper Historians, 1998.

66   Thomas Puttock's Apprenticeship Indenture, Haslemere Museum L.D.5.15.

67   Turner, G., *Shottermill – its Farms, Families and Mills: Part 1 Early Times to the 1700s,* John Owen Smith, Headley, 2004.

68   Turner, G., *Shottermill – its Farms, Families and Mills: Part 2 1730 to the Early Twentieth Century,* John Owen Smith, Headley, 2005.

Index of Paper Mills

The name of each mill, with any alternatives, is followed by its location, including its National Grid Reference, and the river on which it stands. The dates refer to entries in the diaries or the notes and include references which are not specific.

Albury Park, Guildford, Surrey, TQ062479, Tillingbourne. 11 June 1836.

Alton, Hants., SU723395, N. Wey. 12 Feb. 1836; 13 July, 9 Oct. 1837;
    23 Feb. 1838; 26 June 1839; 26 June, 17, 24, 28 Oct., 28 Nov. 1840;
    23 Feb., 20 Oct., 30 Dec. 1841; 7 July, 18 Nov. 1842;
    21, 23 Aug. 1843; 15 Jan. 1852.

Barford Lower, Headley, Hants., SU854380, S. Wey (tributary).
    17 Jan., 27 Feb. 1835; 31 Oct., 16 Dec. 1837; 20 Apr.,
    18 June 1838; 6 Dec. 1849; 23 Jan. 1851; 18 Jan. 1866.

Barford Upper, Headley, Hants., SU854375, S. Wey (tributary).
    27 Feb. 1835; 6 Dec. 1849.

Beech or Lower Marsh, High Wycombe, Bucks., SU891913, Wye
    (tributary). 10 Jan. 1840.

Bowden, High Wycombe, Bucks., SU885922, Wye. 10 Jan. 1840.

Bramshott or Passfield, Hants., SU819345, S. Wey. 17 Jan. 1835;
    11 June 1836; 31 Oct. 1837; 30 July, 16 Aug., 29 Sept.,
    13 Nov. 1841; 7 July 1842; 6 Dec. 1849; 23 Jan. 1851; 20 June 1854.

Carshalton, Surrey, TQ281648, W. Wandle. 4 July 1838.

Catteshall, Godalming, Surrey, SU982443, Wey. 2 Sept. 1831;
    12 Feb. 1841; 4 Sept. 1847.

Chilworth, Guildford, Surrey, TQ024475, Tillingbourne. 9 June 1837;
    13 May 1839; 16 June 1841; 15 Jan. 1852.

Clapton, Wooburn, Bucks., SU910900, Wye. 9, 10 Jan. 1840.

Dalmore, Penicuik, Midlothian, NT252616, N. Esk. 18 Nov. 1848.

Eashing, Godalming, Surrey, SU946437, Wey. 3 Feb. 1835; 16,
17 Nov. 1852.

Fountains, Bermondsey, Surrey, TQ345796, steam. 3 June, 16, 22,
24 Sept., 6, 7 Oct. 1840.

Frogmore, Hemel Hempstead, Herts., TL058055, Gade. 25 Aug. 1845.

Fullers or Gunpowder, Wooburn, Bucks., SU898872, Wye.
7, 8, 10 Jan. 1840; 7 July, 18 Oct. 1841.

Glory or Upper Glory, Wooburn, Bucks., SU912895, Wye.
9, 10 Jan. 1840; 21 Aug., 25 Nov. 1843; 12, 13 Mar. 1845;
14 Apr. 1851; 15 Jan. 1852; 28 Nov. 1853; 8 Aug. 1854.

Golden Valley, Bitton, Somerset, ST682698, Boyd. 18 Nov. 1848.

Grange, Bermondsey, Surrey, TQ346788, steam. 3 Feb. 1835.

Hedge, Loudwater, Bucks., SU907900, Wye. 9, 10 Jan. 1840;
21 Aug. 1843; 14 Apr. 1851; 28 Nov. 1853.

Hedsor, Wooburn, Bucks., SU896869, Wye. 7, 8, 10 Jan.,
18 Oct. 1840; 7 July 1841; 25 May 1848; 30 Nov. 1853.

Iping, near Midhurst, Sussex, SU853229, Rother. 3 Feb.,
11, 13 July 1835; 19 Feb. 1836; 1 Apr., 9 Dec. 1839; 22 Jan.,
1 Nov. 1840; 22 Sept. 1841; 17 May 1848; 16 Nov. 1852.

Ivy, Maidstone, Kent, TQ755525, Loose. 13 May 1839.

Ivy House, Hanley, Stoke on Trent, SJ893473, steam. 13 Mar. 1851.

Kings or New, near High Wycombe, Bucks., SU899912, Wye.
10 Jan. 1840.

Laverstoke, near Whitchurch, Hants., SU492487, Test. 11 June 1836.

Loudwater, Bucks., SU902908, Wye. 10 Jan. 1840.

Lower Glory or Town, Wooburn, Bucks., SU915892, Wye.
9, 10 Jan. 1840; 12 Mar., 24 May 1845; 20 Oct. 1851.

Lower Tovil, Maidstone, Kent, TQ754544, Loose. 13 May 1839.

Marsh Green, High Wycombe, Bucks., SU877921, Wye (tributary).
10 Jan. 1840.

New or Hall's, Linchmere, Sussex, near Haslemere, Surrey, SU881324,
S. Wey. *Passim.*

Pan or Pann, High Wycombe, Bucks., SU869928, Wye. 10 Jan. 1840.

Pitfold, Frensham, near Haslemere, Surrey, SU881327, S. Wey
(tributary). *Passim.*

Postford, Albury, near Guildford, Surrey, TQ039480, Tillingbourne.
11 June 1836; 24 Nov. 1838; 26 Dec. 1839.

Rye or New, High Wycombe, Bucks., SU874926, Wye. 10 Jan. 1840.

St Catherine or Yeeles, near Bath, Somerset, ST782697, St Catherine's
Brook. 17, 24 Dec. 1842.

St Mary Cray, near Bromley, Kent, TQ.472684, Cray. 6 Sept. 1862.

Sickle, Haslemere, Surrey, SU888324, S. Wey. *Passim.*

Soho, Wooburn, Bucks., SU908877, Wye. 10 Jan. 1840.

Standford, Headley, Hants., SU813350, S. Wey. 17 Jan. 1835;
23 Jan. 1851.

Stoke, Guildford, Surrey, SU998510, Wey. 3 Feb. 1835; 24 Nov. 1838.

Test, Romsey, Hants., SU348215, Test. 12 Feb. 1836.

Thames, Bourne End or Lower, Wooburn, Bucks., SU895865, Wye.
10 Jan. 1840.

Trevarno, near Bath, Somerset, ST790672, Avon. 17 Dec. 1842.

Turkey, Maidstone, Kent, TQ772554, Len. 26 June 1839.

Two Waters, Hemel Hempstead, Herts., TL055057, Gade.
4 Mar. 1846.

Up or West End, South Stoneham, Southampton, Hants., SU454156,
Itchen. 12 Feb. 1836.

Wandsworth Royal, Surrey, TQ257741, Wandle. 15 Jan. 1852.

West Ashling, near Chichester, Sussex, SU807074, Ratham.
8 Dec. 1841; 15 Apr. 1852.

Westbrook, Godalming, Surrey, SU967442, Wey. 24 Nov. 1838.

Woking, Surrey, TQ015565, Wey. 2 July 1835; 13 May 1839;
12 Mar. 1853.

Wookey Hole Lower, near Wells, Somerset, ST531465, Axe.
17, 24 Dec. 1842; 30 Jan. 1843.

Wookey Hole Upper, near Wells, Somerset, ST532479, Axe.
20 Oct. 1848.

Index of Personal Names

The biographical notes provided here are based mainly on information collected from parish registers, census returns, local and national directories, the Cooper archives,[8] Swanton's *Bygone Haslemere,*[6] *The Simmons Water Mills Collection,*[5] and of course the diaries themselves. In many cases more specific references are given in the notes on the extracts. Dates refer to entries in the diaries, those with an asterisk being presented in the Introduction or in figure captions.

Acton, Mrs (1769-1848). Wealthy relative of Ann, wife of James IV. 18, 25, 30 May 1848.

Albert, Prince (1819-61). Consort of Queen Victoria. 10 Feb. 1840; 21 Mar. 1860; 15 Dec. 1861.*

Allnutt, snr., Henry. Master paper-maker of Lower Tovil and Ivy Mills, Maidstone, Kent. 13 May 1839.

Allnutt, jnr., Henry (d. 1879). Master paper-maker of Woking, Chilworth and Maidstone Mills. 13 May 1839.

Andrews, Charles. Engineer employed by James III. 22 Apr. 1841.

Angell, Caroline. Master paper-maker of Thames, also known as Bourne End or Lower Mill, Wooburn, Bucks. 10 Jan. 1840.

Appleton, Daniel (d. 1790). Paper-maker from Pitfold. 16 Oct. 1854.

Appleton, Henry (1789-1871). Braid manufacturer of London and at Elstead and Haslemere Mills. 15, 18, 24, 30 Apr., 1, 22-23 May, 4, 7 July 1835; 2 May 1837; 16, 23 Oct., 12 Dec. 1854; 18 Nov. 1857; 23 Oct. 1858; 13 Sept. 1860.

Appleton, John. Paper-maker near Manchester. 16 Oct. 1854.

Appleton, Susannah (1819-1894). Daughter of Henry; married to Samuel Pewtress. 17 Oct. 1840; 16 Oct. 1854; 23 Oct. 1858.

Appleton, Thomas Giles (1815-1899). Son of Henry paper and braid manufacturer of Elstead and Haslemere Mills. 15, 18 Apr. 1835; 16, 23 Oct. 1854.

Appleton, William (1821-1898). Son of Henry and paper and braid manufacturer of Elstead and Sickle Mills. 16, 18, 19, 23 Oct. 1854.

Arnott, Dr. Neil (1788-1874). Inventor of a smokeless grate. 8 Dec. 1838.

Atkins, Absolam. Foreman or clerk working for James III. 14 Mar., 13 Apr. 1840.

Aylwyn, William (1794-1872). Carpenter of Haslemere. 14 June 1847.

Bailey, Henry Virtue (1808-79). Master paper-maker of Woking Mill. 12 Mar. 1853.

Baker, Rev. Richard Henry (1785-1849). Perpetual curate and patron of Linchmere; owner of Shulbrede Priory. 24 Mar. 1836; 23 Mar. 1840.

Baker, Edward James (1782-1844). Farmer of Frensham Hall. 19 May 1836.

Ball, Charles (d. 1820). Master paper-maker of Albury Park Mill. 11 June 1836.

Balston, Messrs. Hollingworth &. Wholesale stationers of London and master papermakers of Turkey Mill and other mills near Maidstone, Kent. 26 June 1839.

Barker, John. Engineer employed by James IV. 25 Feb. 1854.

Barlow, Elizabeth. Acquaintance of James III at Midhurst in 1790s. 21 Aug. 1844.

Batchelor, Rev. Thomas (b. 1799). Curate of Haslemere. 30 Apr. 1842.*

Bertram, George, William and James. Manufacturers of paper-making machines in Scotland. 24 Aug. 1840.

Bowles, Rev. F.A. (b. 1814). From Singleton, Sussex. 31 Mar. 1842.*

Boxall, George (b. 1806). Agricultural labourer employed by James III. 12 Aug. 1841.

Bridger, Christopher. Workman employed by James IV. 31 Jan., 9 Dec. 1853.

Brookman, William. Master paper-maker of Test Mill, Romsey, Hants. 12 Feb. 1836.

Burnside, Messrs. Hodgkinson &. Wholesale stationers of London. 20 Oct. 1848.

Candy, Rev. Charles (1800-90). Vicar of Shottermill. 2 May 1848;*
5 June 1850; 26 May 1858.

Carey, Rev. John Henry Spelman (1771-1852). Vicar of Fernhurst.
28 Sept. 1840.

Carruthers, James (b. 1801). Millwright of Godalming. 2 Oct.,
29 Dec. 1837; 1 May 1838; 28 Feb., 11 May 1840; 16 Apr. 1841.

Cartwright, Edmund (1743-1823). Inventor of power loom.
29 May 1858.

Chalcraft, Mr. Relative of the Warren family, master paper-makers of
Bramshott. 29 Sept. 1841.

Chorley, Robert. Engineer, millwright and brass and iron founder of
Midhurst 19, 22, 26 Sept., 6, 13, 27, 31 Oct., 11 Nov., 12 Dec. 1840;
16 Feb., 27 Mar., 15 Apr., 7, 8 May, 30 July 1841; 31 May,
15 Nov., 15 Dec. 1843; 31 May 1851; 7, 15 Apr. 1852.

Clothier, Henry (1808-84). Medical practitioner of Haslemere.
9 Dec. 1853; 18 Aug. 1857; 25 Dec. 1867.*

Coles, Joseph. Master paper-maker of Lower Wookey Mill, Wells,
Somerset. 17 Dec. 1842.

Comerford, Micheal (1774-1844), Stationer and bookseller of
Portsmouth. 4 Feb. 1841.

Cooper, Catherine (1814-92). Daughter of James III and married to
William Cooper. 16 Oct. 1845, 14 July 1847, 2 June 1860,
28 Dec. 1861,* 6 Sept. 1862.

Cooper, John Eggar (1850-1941). Son of James III's daughter
Catherine. 21 Apr. 1857.*

Cooper, William (1806-65). Wholesale stationer and wallpaper
manufacturer of London, husband of James III's daughter
Catherine and great-grandfather of William J. Dupre Cooper,
who has made the diaries available. 13, 14 July 1847; 10 Oct. 1851;
6 Sept. 1862.

Cowper, Edward (1790-1852). Inventor of a paper cutting machine.
17 Dec. 1840.

Croft, John (c1791-1869). Paper-maker employed by James III.
12 Aug. 1841.

Curtis, Thomas. Wholesale stationer of London and master paper-
maker of Chesham Bois Mill, Bucks. 9 Nov. 1838; 13 Apr. 1840.

Daintrey, Adrian (b. 1816) and Arthur (b. 1804). Solicitors of
Petworth. 19 Mar. 1839; 21 Apr., 9 Oct. 1840; 14 Nov. 1851;
30 Sept., 6 Oct. 1852.

Dean, William. Suspected thief of Hindhead. 31 May, 5 June 1837.

Dickinson, John (1782-1869). Wholesale stationer, paper-maker and inventor of London and Herts. 28 Jan. 1836; 9 July 1839.

Donkin, Bryan (1768-1855). Engineer who developed the first paper-making machine. 21 Jan. 1836; 7 Oct., 19 Nov. 1839; 12 Feb. 1841; 7 Mar. 1844; 4 Apr., 25 Aug. 1845; 17 Aug. 1847.

Donkin & Wilks Messrs. Engineers of Dartford and Bermondsey. 24 Aug. 1840.

Dunster, Mr. Advisor of James IV in London. 26 Feb., 18 Mar., 1, 5 Oct. 1853.

Dutton, William. Assistant overseer for Frensham parish. 25 Mar. 1845.*

Edmonds, Thomas. Master paper-maker of Pan Mill and Rye Mill, also called New Mill, High Wycombe, Bucks. 10 Jan. 1840.

Edmonds, Messrs. Lane &. Paper-makers of Marsh Green and Bowden Mills, High Wycombe, Bucks. 10 Jan. 1840.

Edmonds, Mr. Dissenting minister from Petworth. 14 Oct. 1839.*

Eliot, George (Mary Ann Evans) (1819-80). Novelist who stayed at Brookbank, Shottermill in 1871. 3 Jan. 1852.

Elliott, Rev. Laurence William (1777-1862). Rector of Peper Harow. 31 May 1837.

Everitt, Samuel Philip. Tenant paper-maker of James III at Pitfold Mill. 30 Aug., 1, 7 Sept. 1831.

Ewen, Alethea. Wife of churchwarden of Shottermill. 12 Aug. 1842.

Fairbairn, William (1789-1874). Patentee with John Hetherington of Lancashire boiler. 18 Nov. 1857.

Findley, Mr. Partner of J.L. Lightfoot, tenant of Sickle Mill. 8 June 1849.

Fourdrinier, Henry (1766-1854). Stationer who with his brother Sealy financed development of the paper-making machine. 4 Mar. 1846.

Fourdrinier, George Henry (d. 1869). Brother of Joseph and master paper-maker of Ivy House Mill, Hanley, Stoke on Trent. 13 Mar. 1851.

Fourdrinier, Joseph (1795-1862). Eldest son of Henry and tenant of Sickle Mill. 4 Mar. 1846; 16, 18, 30 Sept., 5, 8 Oct., 9, 16 Nov., 11, 14 Dec. 1850; 4, 8, 11, 18, 21 Jan., 1 Feb., 1, 13, 26 Mar., 12 Apr., 23 Aug., 27 Oct., 26 Nov., 20 Dec. 1851; 3, 10 Jan., 27 Mar., 19 Apr.,* 22 May, 31 July, 16, 31 Aug., 2, 6, 18, 21, 30 Sept., 6, 8, 13, 14, 16, 18, 23 Oct., 11 Nov., 2 Dec. 1852; 5 Feb. 1853.

Fourdrinier, Sophia, *née* Brooks, (1807-68). Wife of Joseph.
6, 12 Oct. 1852.

Fox, Edward (1798-1873). Silk mercer of Snaresbrook, Wanstead,
Essex. 21 Oct. 1841; 7 Mar. 1844.

Fox, Emma, *née* Lunnon (1819-1881). Wife of Edward and elder sister
of Ann who married James IV. 21 Oct. 1841; 7 Mar. 1844;
25 May 1848.

Fromow, Peter John. Master paper-maker of Clapton Mills, Wooburn,
Bucks. 9, 10 Jan. 1840.

Fryer, Michael L. Miller of Mill End, West Wycombe, Bucks.
10 Jan. 1840.

Gater, John or Edward. Master paper-makers of Up Mill, South
Stoneham, Hants. 12 Feb. 1836.

Gaviller, Augustine. Master paper-maker of Loudwater, near
Wooburn, Bucks. 10 Jan. 1840.

'George'. Carter employed by James III. 5 Nov. 1838; 20 May 1839;
6 Oct. 1840; 18 Jan. 1841.

Gibson, Thomas-Milner (1806-84). President of Board of Trade.
20 Feb. 1860.

Gladstone, William Ewart (1809-98). Chancellor of Exchequer.
20 Feb. 1860.

Goddard, Mr. Bow Street Officer. 4, 5, 6 June 1837.

Goovenn [Gover?] & Co. Wholesale stationers of London.
5 Aug. 1841.

Gordon, George Frederick (c1770-1852). Surgeon of Haslemere.
11 June 1839.

Greenfield, George and Sarah. Grocers, drapers and rag dealers of
Storrington, Sussex. 27 July 1836.

Harding, Abraham. Master paper-maker at Barford Mills. 27 Feb. 1835.

Harding, Abraham (1783-1854). Employee of James III. 29 May 1837;
23 Mar. 1843.

Harding, Abraham Ebenezer (1828-85). Haslemere paper-maker who
emigrated to America. 14 Dec. 1850.

Harding, Charles (b. 1823). Apprentice of James III. Emigrated to
America with his brother Abraham Ebenezer. 19 Mar.,
5, 7 Dec. 1839; 30 Jan. 1840; 18 , 30 Sept. 1841; 20 Feb. 1843.

Harding, James. Paper-maker employed by James III. 27 Feb. 1835.

Harding, James. Carpenter (c1775-1861) at Lion Common near Sickle Mill. 29 Sept. 1836.

Harding, Jeremiah. Corn miller of Shotter Mill. 28 Sept. 1839.

Harding, John (1814-87). Paper-maker employed by James III. 27 Feb. 1835; 1 Dec. 1837.

Harris, Sir William (1807-48). Bombay engineer and African traveller who visited Sickle Mill with Lady Harris to see the machine. 12 May 1845.

Hearne, Mr. Shopkeeper of Wycombe, Bucks. 13 Mar. 1845.

Hetherington, John. Patentee with William Fairbairn of the Lancashire boiler. 18 Nov. 1857.

Hewitt, William (1795-1843). Relative of James III at Vauxhall. 1 Aug. 1840.

Hickman & Marriott, Messrs. Wholesale stationers of London. 2 July 1835.

Higginbottom, Newburgh. Deputy Surveyor of Alice Holt Forest and supplier of wood for boiler at Sickle Mill. 2 Apr. 1853.

Hill, Rowland (1795-1879). Proposer of universal penny post. 11 Jan. 1840.

Hodgkinson & Burnside, Messrs. (see Burnside).

Hodgkinson, William Sampson. London stationer and master paper-maker of Upper Mill, Wookey Hole, Somerset. 20 Oct. 1848.

Hollingworth & Balston, Messrs. (see Balston).

Howard, Miss. Companion of Mrs Acton. 18 May 1848.

Howard, snr., John. Tenant of Haslemere paper mills, 1801-11 who married Rebecca, sister of Thomas Lunnon, snr., of Wooburn in 1802. 9 Jan., 14 Mar. 1840; 28 Oct. 1843.

Howard, jnr., John. Son of John, snr., and paper-maker near Wooburn and later in Russia. 8, 9 Jan. 1840; 23, 28 Oct. 1843; 18 May 1848.

Hunt & Co. Wholesale stationers of London (J.B. Hunt, Henry Fourdrinier & Sealy Fourdrinier). 12 Nov. 1842.

Hutton, Mr. Sheriff's Officer. 30 Sept. 1852.

James, Mrs. (see Ann Simmons, *née* Lunnon).

Joynson, William. Master paper-maker of St Mary Cray, Kent. 6 Sept. 1862.

Kidd, Mr. Banker of Godalming. 17 Aug. 1847; 22 Apr. 1857.*

Lacy, Mr. Prospective tenant of Sickle Mill. 7 June 1848.

Lane, Mr. Master paper-maker of two mills near Wooburn, Bucks. 10 Jan. 1840.

Lane & Edmonds, Messrs. (see Edmonds, Messrs. Lane &).

Lightfoot, John Lill. Master paper-maker and tenant of Sickle Mill.
5, 12, 15, 19, 24 May, 2 Aug., 6 Dec. 1849; 19 Jan., 9, 15, 30 Mar.,
13, 26, 27 Apr., 4 May 1850.

Lightfoot, Richard. Wire merchant of Dublin. 5 May 1849.

Lightfoot, snr., Mr. Father of John Lill. 5, 12 May, 8 June,
12, 21, 28 July 1849; 15 Mar., 13 Apr. 1850.

Lomas, Mr H. V. Auctioneer, appraiser and Sheriff's Officer of
Guildford. 20, 22 Apr. 1850.

Long, Robert. Rag and rope dealer of Portsmouth. 5 June, 6 July 1837;
4 Feb. 1841; 27 Mar., 20, 22 Apr. 1850.

Lowe, James. Partner of Pewtress brothers at Iping and other paper
mills. 16 Nov. 1852.

Lucas, John (1798-1871). Miller of Lowder Mill, Haslemere. 4 Aug. 1845.

Lucas, C. Son of John of Lowder Mill, Haslemere. 19 June 1837.

Lucas, Mr. Miller of Bedhampton near Portsmouth. 17 Nov. 1840.

Lucas, Mr. Prospective tenant of Sickle Mill, from Bucks. 7 June 1848.

Lunnon, snr., Thomas (1778-1841). Master paper-maker and miller of
Hedsor and Fullers Mills, near Wooburn, Bucks., and father of
Thomas, jnr., Emma Fox and Ann Simmons, wife of James IV.
7, 10 Jan., 24 Aug. 1840; 7 July, 12, 18, 21 Oct. 1841; 16 Oct. 1845.

Lunnon, jnr. (1815-86), Thomas. Master paper-maker trained by James
III and son of Thomas, snr. 10 Jan. 1840; 7 July, 21, 30 Oct.,
13 Nov. 1841; 7 Mar. 1844; 16 Oct. 1845; 25 May 1848; 30 Nov. 1853.

Macdonald, Sir Archibald Keppel (1820-1901). Lord of Ludshott
Manor, equerry to the Duke of Sussex and High Sheriff of Hants.
15 Oct. 1841.*

Mackintosh, Alexander. Proprietor of printing works in London.
7 Mar. 1844.

Magnay, George (d. 1852). Younger brother of William and James,
and master paper-maker of Postford and Stoke Mills near
Guildford. 24 Nov. 1838; 26 Dec. 1839; 7 Mar. 1846.

Magnay, James (1798-1842). Younger brother of William, and master
paper-maker of Postford and Stoke Mills near Guildford and
Westbrook Mill, Godalming. 24 Nov. 1838; 26 Dec. 1839.

Magnay, Jane. Master paper-maker of Postford Mills. 26 Dec. 1839.

Magnay, William (1797-1871). Wholesale stationer and master paper-maker of London. 24 Nov., 15 Dec. 1838; 21 Nov. 1839; 11 May, 17 Dec. 1840; 20 July 1842.

Marriott, Messrs. Hickman &. (see Hickman).

Maybanke, John. Medical gentleman from Wonersh, near Guildford. 30 Jan. 1843.

Mellersh, F. & A. Auctioneers of Godalming. 13 Sept. 1860.

Mellersh, Thomas and Henry. Attorneys of Godalming. 4, 12 Aug. 1845.

Metternich, Clemens Werzel Lothar, Prince von (1773-1859). Austrian statesman. 29 Mar. 1848.

Minor, Edward. Labourer employed by James III. 18 Oct. 1837.

Mintley, Mr. Engineer. 11 Nov. 1843.

Moline, Benjamin or Robert. Bankers of Godalming. 17 Aug. 1847; 22 Apr. 1857.*

Moon, William. Carter of Godalming. 26 Sept., 7 Oct. 1840.

Moorey, John. (1781-1838). Employee of James III. 31 Mar. 1838.

Morbey, Edwin. Master paper-maker of Beech Mill, also called Lower Marsh Mill, High Wycombe, Bucks. 10 Jan. 1840.

Muggeridge, Messrs. Paper-makers of Carshalton. 4 July 1838.

Newman, Ann (1788-1860). Sister of James III and widow of Thomas. 15 Oct. 1841;* 21 Aug. 1844; 25 May 1848; 10 Aug. 1850.

Newman, Anthony (1771-1829). Late husband of James III's sister Catherine; from Cocking, Sussex. 21 Aug. 1844.

Newman, Arthur Thomas (1816-97). Son of James III's sister Ann and farmer of West Dean, Sussex. 31 Dec. 1839.*

Newman, Catherine (1785-1863). Sister of James III and widow of Anthony. 12 Mar. 21 Aug. 1844; 25 May 1848; 10 Aug. 1850.

Newman, George White (1816-55). Son of James III's sister Catherine and miller of Compton windmill and Hurst Mill, South Harting, south of Petersfield. 23 Jan. 1841; 12 Mar. 1844; 24 July 1854.

Newman, John (b. c1824). Son of James III's sister Catherine and apprenticed with Bryan Donkin of Dartford, Kent. 7 Oct. 1839; 7 Mar. 1844.

Newman, Thomas (1784-1836). Husband of James III's sister Ann and farmer of East Marden, Sussex. 21 Aug. 1844.

Newman, White (1821-90). Son of James III's sister Ann and farmer of East Marden, Sussex, and later of Kingsley, Hants. 15 Oct. 1841.*

Noble, James. Patentee of improved wool combing machine. 29 May 1858.

Oliver, George. Miller of Shotter Mill. 28 Sept. 1839; 11 Apr. 1844; 4 Aug., 2 Dec. 1845.

Openshaw, Johnathen (b. 1816). Employee of James IV. 5 Nov. 1853.

Parson, George John (b. 1819). Solicitor of Haslemere. 14 June 1852.

Parson, James (1806-43). Attorney of Haslemere. 19 May 1836.

Pegg, William. Master paper-maker of Wooburn Mill, Bucks. 10 Jan. 1840.

Penfold, Charles. Rag dealer of Arundel. 26 July 1836.

Penfold, snr., John Wornham (1789-1873). Cousin and brother-in-law of James III. 11 June 1839; 10 Aug. 1850; 8 Aug. 1854.

Penfold, jnr., John Wornham (1828-1909). Architect, surveyor and local historian. 2 June 1838; 11 June 1839; 5 June 1847; 10 Aug. 1850.

Penfold, Mary (1790-1872). Sister of James III and wife of John Wornham, snr., 10 Aug. 1850.

Pewtress, Benjamin (1790-1854) and Thomas (1794-1872). Master paper-makers of Iping near Midhurst, Eashing near Godalming, Stoke near Guildford and Bermondsey. 3 Feb., 13 July 1835; 19 Feb. 1836; 1 Apr. 1839; 22 Jan., 1 Nov. 1840; 22 Sept. 1841; 17 May 1848; 10 Aug. 1850; 16, 17 Nov. 1852; 23 Oct. 1858.

Pewtress, Samuel (1818-58). Son of Benjamin or Thomas; married to Susan, daughter of Henry Appleton. 17 Oct. 1840; 23 Oct. 1858.

Pimm, Mr. London agent of John Lill Lightfoot. 20, 26 Apr. 1850.

Plaistowe, Richard. Master paper-maker of Kings Mill, High Wycombe and Loudwater Mill, Bucks. 10 Jan. 1840.

Portal family. Master paper-makers of Laverstoke Mill, Hants. 11 June 1836.

Puttick, Elias (b. 1821). Apprentice of James III. 30 June 1835; 18 July 1839; 3 May 1841.

Puttick, Joseph (c1826-45). Apprentice of James III. 19 Mar., 5, 7 Dec. 1839; 30 Jan., 3 Feb. 1840.

Puttick, Robert (1809-87). Haslemere mechanic employed by James III and James IV. 31 Oct., 16 Dec. 1837; 15 Mar., 13 Oct., 12 Nov. 1838; 1 Dec. 1840; 10 June 1841; 12 Mar. 1853.

Rhoda. Servant of James III. 25 Dec. 1867.*

Ridge & Co. Bankers of Chichester. 1 Dec. 1841.

Roker (Roakes), Henry (b. 1801). Miller of Hatch Mill, Godalming.
8 May 1835; 12 Mar. 1844.

Salmon, Isaac (b. 1820). Apprentice of James III. 30 June 1835,
18 July 1839; 15 Jan. 1842.

Salmon, Robert. Paper-maker employed by James III. 28 Sept.,
1 Nov. 1840; 17 May 1848.

Savage, William, Henry and Soloman. Master paper-makers of West
Ashling Mill near Chichester. 8 Dec. 1841.

Shaw, William. Defendant in court case. 4 Aug., 2 Dec. 1845.

Simmonds, William (1810-92). Miller of Bourne Mill, Farnham.
7 Apr. 1841.

Simmons, Ann (1817-84). Daughter of James III. 15 Oct. 1841;*
2 June 1860; 28 Dec. 1861;* 6 Sept. 1862; 2 June 1864;* 25 Dec. 1867.*

Simmons, Ann (1823-90). Wife of James IV and daughter of Thomas
Lunnon, snr. 7 Jan. 1840; 16 Oct. 1845; 18, 25 May 1848.

Simmons, Catherine. See Cooper, Catherine.

Simmons, Charlotte (1778-1860). Wife of James III and sister of
Anthony and Thomas Newman of Cocking, Sussex. 8 May 1835;
27 July 1836; 19 June 1837; 5 May 1842 ;* 21 Aug. 1844;
16 Oct. 1847; 29 May 1858; 19 May, 2 June 1860; 16 Jan. 1868.

Simmons, Charlotte Hannah. See Small, Charlotte Hannah.

Simmons, Hannah, née Philps (1752-1842). Mother of James III.
12 Jan. 1837; 13 May 1838; 15 Oct. 1841;* 30 Apr. 1842.*

Simmons, James II (1738-90). Uncle of James III and master paper-
maker of Haslemere. 14 Feb. 1852.

Simmons, James IV (1815-1903). Elder son of James III and master
paper-maker of Haslemere, who became one of Surrey's first
aldermen. *Passim.*

Simmons, Sarah (1798-1889). Sister of James III. 10 Aug. 1850;
9 Dec. 1854.

Simmons, William (1748-1801). Father of James III and master paper-
maker of Haslemere. 12 Jan. 1837; 9 Jan. 1840; 14 Feb.,
18 Oct. 1852.

Simmons, William (1820-96). Younger son of James III. 21 Jan. 1837;
13 Dec. 1839; 23 Feb., 6 Mar., 13 Nov. 1841; 13 Aug.,
18 Nov. 1842; 17 May 1843; 16 Oct. 1845; 25 May 1848;
20 Oct. 1851; 15, 31 Jan. 1852; 8 Aug.,* 10, 19 Oct. 1854.

Small, Charlotte Hannah, *née* Simmons, (1810-81). Daughter of James III. 19 June 1837; 29 May 1858.

Smith, George (1793-1855). Surgeon of Haslemere. 11 June 1839; 30 Apr. 1842.*

Smith, Sidney. Lawyer of Holborn, London, and partner of Joseph Fourdrinier. 18, 21 Sept., 5 Oct., 9 Nov., 11, 14 Dec. 1850; 13 Feb., 26, 29 Mar. 1851; 31 Aug. 1852.

Sommerville, jnr., William (1807-99). Master paper-maker of Dalmore Mill, Penicuik, near Edinburgh, and later of Golden Valley Mill, Bitton, near Bristol. 18 Nov. 1848.

Spicer, Freeman Gage (1769-1834). Father of James Freeman Gage 21 Aug. 1843.

Spicer, James Freeman Gage (1808-69). Master paper-maker of Glory and Hedge Mills etc. near Wooburn, Bucks. 9, 10 Jan. 1840; 21 Oct. 1841; 21, 23 Aug., 25 Nov. 1843; 12, 13 Mar. 1845; 14 Apr., 20 Oct. 1851; 15, 31 Jan. 1852; 28 Nov. 1853; 8 Aug. 1854.

Spicer, snr., John Edward (1766-1845) . Master paper-maker of Alton. 12 Feb. 1896; 21 Aug. 1849.

Spicer, jnr., John Edward (1803-69). Son of John Edward, snr., and master papermaker of Alton. 12 Feb. 1836; 13 July, 9 Sept., 9 Oct. 1837; 23 Feb. 1838; 26 June 1839; 26 June, 11 July, 17, 24, 28 Oct., 28 Nov. 1840; 23 Jan., 23 Feb., 6 Mar., 20 Oct., 30 Dec. 1841; 7 July, 18 Nov. 1842; 15 Jan. 1852.

Stedman, Richard (1782-1857). Bookseller, wine merchant, clockmaker and insurance agent of Godalming and cousin of James III. 10 July 1835; 28 Feb., 19 Mar. 1840.

Sumner, Charles Richard (1790-1874). Bishop of Winchester. 22 Mar. 1841; 2 May 1848.*

Sweetapple, Thomas (1793-1870). Master paper-maker of Catteshall Mill, Godalming. 2 Sept. 1831; 12 Feb. 1841; 4 Sept. 1847.

Taylor, Sir Charles William (1770-1857). Magistrate and owner of Hollycombe estate. 5 Dec. 1839.

Teasdale, James (b. 1801). Builder of Haslemere. 28 Oct. 1840; 14 June 1847.

Tennyson, Alfred (1809-92). Poet who toured Haslemere with James IV. 3 Jan. 1852.

Tilbury, John (1781-1852). Paper-maker at Haslemere mills. 14 Feb. 1852.

Tilbury, Mrs. Neighbour of James III who turned sceptic. 30 Aug. 1848.*

Timms, George. Boy working at Haslemere mills. 11 June 1839.

Trimmer, Charles (1791-1848). Owner of Anstead Brook, Chiddingfold. 12 May 1845.

Turner, George William. Master paper-maker of Fountains Mill, Bermondsey. 3 June, 16, 24 Sept. 1840.

Tyrrell, Edward (1792-1881). City of London Remembrancer and London adviser of Lunnon family. 7 July 1841.

Venables, snr., Charles. Master paper-maker of Lower Glory Mills, Wooburn, Bucks. 9, 10 Jan. 1840.

Venables, jnr., Charles. Master paper-maker of Soho Mill, Wooburn, Bucks. 10 Jan. 1840.

Venables, Francis Edward. Master paper-maker of Fullers and Lower Glory Mills, Wooburn, Bucks. 10 Jan. 1840; 20 Oct. 1851.

Venables, George Henry (b. 1824). Master paper-maker of Lower Glory and Clapton Mills, Wooburn, Bucks. 10 Jan. 1840; 20 Oct. 1851.

Venables, Alderman William (1776-1840). Wholesale stationer of London and papermaker of Bucks. and Woking. 2 July 1835; 9 July 1836; 13 May 1839; 1 Aug. 1840.

Verstage, Mr. Flock-maker of Barford Lower Mill. 18 Jan. 1866.

Victoria, Queen (1819-1901). 28 June 1838,* 10 Feb. 1840; 21 Mar. 1860.

Voller, snr., William. Paper-maker apprenticed to William Simmons in 1800. 18 Oct. 1852.

Voller, jnr., William. Employee of James III and James IV. 18 Oct., 26 Nov. 1852.

Walther, Mr. Bookseller of Piccadilly, London. 3 Feb. 1843.

Ward, Mr. Grocer and rag dealer from Beccles, Suffolk. 21 Mar. 1835.

Warren, Andrew (c1821-99). Son of William and master paper-maker of Bramshott, Lower Barford and Standford Mills. 15 Nov. 1841; 18 Jan. 1866.

Warren, George Roe (1814-92). Son of William and master paper-maker of Bramshott, Lower Barford and Standford Mills. 6 Mar., 15 Nov. 1841; 2 Mar., 7 July 1842; 18 Jan. 1866.

Warren, William (1787-1860). Master paper-maker of Bramshott, Lower Barford and Standford Mills. 17 Jan. 1835; 12 Feb., 11 June 1836; 31 Oct. 1837; 30 July, 16 Aug., 27, 29 Sept., 13 Nov. 1841; 7 July 1842; 23 Jan. 1851; 18 Jan. 1866.

Webb, Col. Robert. Magistrate of Milford near Godalming. 5 Dec. 1839.

Wellington, Arthur Wellesey, Duke of (1769-1852). Soldier and statesman. 29 Mar. 1848.

West, James. Master paper-maker of Standford Mill. Headley. 17 Jan. 1835.

Wilberforce, Rev. Samuel (1805-73). Archdeacon of Surrey who became Bishop of Oxford. 31 Mar. 1842.*

Wilks, Messrs. Donkin &. (See Donkin).

Wilson, Dr. George (1818-59). Scientist from Edinburgh. 18 Nov. 1848.

Winchester, Mr. Suspected thief. 5 June 1837.

Wright, Samuel Newell. Master paper-maker of Wooburn, Bucks. 10 Jan. 1840.

Wright, Sarah. Married John Edward Spicer, jnr., of Alton. 7 July 1842.

Wyatt, Edward (1805-90). Farmer of Up Marden south of Petersfield and married to Ellen Newman, niece of Charlotte Simmons. 12 Mar. 1844.

Yeeles, George. Master paper-maker of Trevarno Mills, near Bath, Somerset. 17 Dec. 1842.

Yeeles, Robert. Master paper-maker in 1626 of St Catherine's or Yeeles Mill near Bath, Somerset. 17, 24 Dec. 1842.

The Authors

Alan Crocker is a Professor Emeritus in the Physics Department at the University of Surrey and President of the Surrey Industrial History Group. He is also a Past-President of the Surrey Archaeological Society and a Past-Chairman of the British Association of Paper Historians, the Gunpowder Mills Study Group and the Mills Section of the Society for the Protection of Ancient Buildings. His publications include several books including *Catteshall Mill* with his wife Glenys in 1981, about a mill in Godalming that made paper from 1661 to 1928, *Paper Mills of the Tillingbourne* in 1988 about five mills on a tributary of the River Wey, the first edition of *The Diaries of James Simmons* with Martin Kane in 1990, *Damnable Inventions* with Glenys in 2000, the title being a comment made by William Cobbett in his book *Rural Rides* to describe the production of paper and gunpowder at Chilworth. He has also published about 100 shorter articles on aspects of industrial archaeology and about 150 scientific research papers.

Martin Kane is a Surrey born local historian and writer with a special interest in the history of Haslemere. In the late 1980s he worked as historian for the Liphook based River Wey Trust. This Trust was set up to carry out conservation work and to research the past use of the southern branch of the river between Haslemere and Tilford. His research was published in *The Southern Wey*, a guidebook produced by the Trust in 1988. This work led directly to the discovery of James Simmons's diaries. In addition to collaborating with Alan Crocker he edited *Who's Who in the Diaries of James Simmons*, to accompany a full transcript which was completed by members of Haslemere Local History Group in 1990. *A Country Museum Revisited* followed in 1995, a history of the first hundred years of Haslemere Educational Museum. A long-term volunteer at the Museum, Martin was for several years an honorary archivist and assisted in cataloguing new acquisitions, mounting exhibitions on local themes such as Penfold's Pillar Box and responding to local history queries on behalf of the Museum.